At the crossroads
of cultures

In this series:

Cultures and time
Time and the philosophies
Time and the sciences

In preparation:

Cultures and history

TIME and the SCIENCES

J. M. Abraham   Ahmed M. Abou-Zeid
F. N'Sougan Agblemagnon   Paul Fraisse
Mübeccel Kiray   John McHale   A. Neelameghan
S. L. Piotrowski   Frederic Vester   Gerald J. Whitrow

Edited by Frank Greenaway

415

Published in 1979 by the United Nations
Educational, Scientific and Cultural Organization
7 Place de Fontenoy, 75700 Paris
Photosetting by Thomson Press (India) Limited, New Delhi, and
printed by Imprimerie de la Manutention, Mayenne

ISBN 92-3-101657-1
French edition: 92-3-201657-5
Spanish edition: 92-3-301657-9

PREFACE

These essays constitute the third of a series forming part of a Unesco project under the title 'At the Crossroads of Cultures', aimed at establishing communication between contemporary cultures through studies of fundamental concepts that are not only immediately evident in daily life but are central also to any form of systematic and comprehensive view of the world. Science, pure and applied, derives from the ordering of men's thoughts and from the orderly control of action upon the objects of the surrounding world. In reducing thought and order to action, the experience of time plays a leading role, being one of the elements of our experience of conscious existence. Of those elements, it shares with space a susceptibility to measurement and quantitative expression, but nevertheless has features that challenge our concepts of quantitative thought. The concept of time of the rural peasant has complexities and subtleties of its own, and is no less puzzling and deserving of study than that of the sophisticated scientist.

It is therefore important to draw together the concepts of time that are developed in various fields of organized study, from the mathematical to the sociological. The essays collected here are based on papers presented at an expert meeting on 'Time and the Sciences: Impact of Scientific Expressions of Time on Differing Cultures', held at the Royal Institution in London in February 1976. The meeting was organized by the International Union of the History and Philosophy of Science on behalf of the International Council of Scientific Unions and under the auspices of the Philosophy Division of Unesco. One thing above all else emerged at that meeting: namely that, however remote one avenue of scientific inquiry may seem to be from another, the study of its time elements reminds us all that there is a common humanity of which science, in whatever form, is a universal expression.

The authors are responsible for the choice and the presentation of the facts contained in this book and for the opinions expressed therein, which are not necessarily those of Unesco and do not commit the Organization. The designations employed and the presentation of material throughout the publication do not imply the expression of any opinion whatsoever on the part of Unesco concerning the legal status of any country, territory, city or area or of its authorities, or concerning the delimitation of its frontiers or boundaries.

CONTENTS

FOREWORD

Sir George Porter, F.R.S.

Sir George Porter, F.R.S., Director of the Royal Institution of Great Britain, Nobel Prize winner in Chemistry for 1967, gave the following opening address at the meeting on 'Time and the Sciences', held from 4 to 6 February 1976, at the Royal Institution of Great Britain, in London.

Time is the common experience of man, but it is also a factor in that detailed examination of the world which, above all other types of investigation, has become characterized by the name of science.

Scientific method has made three major contributions to our concept of time. First, it has given a definition to the direction of time through the second law of thermodynamics and its statistical mechanical interpretation; secondly, it has revealed the inseparable connection between time and space and the relativity of time measurements from one observer to another; and, thirdly, it has shown the limits of certainty in measurements of quantities made in finite intervals of time. We can hardly claim that, as a result of these discoveries, we understand better the meaning of time; on the contrary, they have merely emphasized how very far we are from any such understanding.

On the other hand, in the experimental sciences, the conquest of time has made spectacular progress. The precision of measurement of time intervals greatly exceeds that of length or mass. The study of fast chemical change and of substances of transient existence has been the main theme of our research here at the Royal Institution for the last decade, and during that period the time intervals which can be resolved have been reduced by a factor of a million from a microsecond to a picosecond (10^{-12}s). Another factor of one thousand will take us to a femtosecond and the uncertainty limit of time as far as chemistry is concerned. At shorter times than this it can be said that chemistry no longer exists.

Once one deals with questions of time, even in photochemistry, one quickly reaches provocative philosophical depths of great interest. What may seem to the plain man to be simple things, i.e. one's experience of time and space, are in fact the most difficult obstacles to our comprehension of the world. In his search for knowledge man has met many impedimenta in the form of apparently unknowable things. But it is precisely the pursuit of the unknowable that provides the driving force for science and enables it to relate to other aspects of human culture. The researches being carried out at the Royal Institution characterize this wider role of science, since the Royal Institution, like Unesco, is an educational, scientific and cultural organization. It is, therefore, an entirely suitable place for a meeting, the conclusions of which may have implications beyond the bounds of science itself.

INTRODUCTION

Gerald J. Whitrow

Until the rise of modern industrial civilization men's lives were far less consciously dominated by time than they have been since. In the last two hundred or so years, the concept of time has also come to play an increasingly important role in the sciences. Although the significance that we attach to the related concepts of time and history makes it seem both obvious and inevitable that, not only our own lives, but the workings of nature, too, should be dominated by them, in the past men thought very differently. It is natural to ask why time has become so important for us. Is it an automatic consequence of the industrial revolution, or is there some other explanation? The development of the mechanical clock and, more recently, of portable watches, has certainly had a tremendous effect on our lives, but there have also been other influences that have affected our concept of time, notably the way in which the various sciences have developed, and the discoveries that have been made about the temporal properties of natural phenomena and of the world generally. The essays here collected, which are based on papers delivered at the Unesco conference on 'Time and the Sciences', held at the Royal Institution, London, on 4–6 February 1976, aim at describing and assessing the role of time in the various sciences, and the effects on society generally of changing ideas about its nature and significance.

Time and mankind

A sense of time involves some awareness not only of duration but also of the differences between past, present and future. There is evidence that our sense of these distinctions is one of the most important mental faculties distinguishing man from all other living creatures. For we have good reason to believe that all animals except man live in a continual present. The possession by animals of some sense of memory, as shown, for example, by dogs who are inclined to give vent to the wildest joy on seeing their masters after long separation, does not necessitate any image of the past as such. It is sufficient for the dog to recognize its master. The point has been put very forcibly by Ernst Cassirer. Arguing that in man we cannot adequately describe recollection as the mere return of a faint image or copy of former impressions, he claims that it is a constructive process. 'It is not enough', he writes, 'to pick up isolated data of our past experience; we must really re-collect them, we must organize and synthesize them, and assemble them into a focus of thought. It is this kind of recollection which gives us the characteristic human shape of memory and distinguishes it from all other phenomena in animal or organic life.'[1]

Similarly, there is no firm evidence that animals have any sense of the future as such. In general, any actions of theirs that might be thought to bear on this question seem to be purely instinctive, although this conclusion is not quite so obvious in the case of the higher apes, particularly the chimpanzee. The problem was examined very carefully by Wolfgang Koehler in the course of his famous investigation of the mentality of apes.[2] He studied cases where chimpanzees undertook, with a view to some final goal, preparatory work that lasted a long time and in itself afforded no visible approach to the desired end. In such cases it seemed at first that the animal might have some rudimentary notion of the future. Nevertheless, Koehler came to the conclusion that all such behaviour by the highest apes could be explained more directly from a consideration of the present only. In particular, after a careful analysis of experiments in which chimpanzees readily responded to the opportunity given them to postpone eating until they had accumulated a large supply of food to eat later in some quiet corner free from disturbance, Koehler could find no reason for interpreting their conduct as evidence for a sense of the future. Instead of the animal being spurred on by some feeling of what it will be like later on when he eats the food, he believed that the chimpanzee's behaviour was simply a response to its instinctive desire to get as much food as possible now.

The conclusion that a sense of time is a peculiar characteristic of mankind needs careful consideration. For whereas in the absence of any incontrovertible counter-evidence we have good reason to deny this faculty to animals, it has been claimed that there are human beings who manage very well without it. The example that is often cited is that of the Hopis of Arizona, whose language was studied in great detail by Benjamin Lee Whorf.[3] He came to the conclusion that the Hopi language contains no words, grammatical forms, constructions or expressions that refer to time or any of its aspects. Instead of the concepts of space and time, the Hopi uses two other basic forms, which Whorf denoted by the terms objective or manifested, and subjective or manifesting. The objective or manifested comprises all that is or has been accessible to the senses, with no distinction being made between present and past, although everything that we call future is excluded. The subjective or manifesting comprises all that for us is future, much of which the Hopi regards as predestined, at least in essence. It is the realm of expectancy, desire and purpose, and can be regarded as a dynamic state that not only embraces what we regard as future but also includes an aspect of the present, namely that which is beginning to be manifested or done, like starting an action such as going to sleep. Whereas the objective state includes all extensional aspects of existence, all intervals and distances and in particular the temporal relations between events that have already happened, the subjective state comprises nothing corresponding to the sequences and successions that we find in the objective state. Although the Hopi language prefers verbs to nouns, its verbs have no tenses. Moreover, the Hopi do not need terms that refer to space or time. Terms that we use to refer to these concepts are replaced by expressions concerning extension, operation and cyclic process if they refer to the objective realm. Terms that refer to the future, the psychic-mental, the mythical and the conjectural are replaced by expressions of subjectivity. Whorf claimed that, as a result, the Hopi language gets along perfectly without tenses for its verbs.

Whorf's contention that the Hopi language contains no reference to time, either explicit or implicit, is, however, too sweeping. For there is a temporal distinction between the two basic forms of Hopi thought. Instead of the three temporal states that we explicitly recognize—past, present and future—the Hopi imagines two distinct states which, between them, comprise our past, present and future. In so far as the Hopi recognizes implicitly a distinction between past and future, he cannot be said to live only in the present. He has some sense of time, although his fundamental intuition of time is not the same as ours.

Whorf's investigation of the Hopi language therefore reveals that, just as our intuition of space is not unique—for since the discovery last century of non-Euclidean geometries we now know that there is no unique geometry that must necessarily apply to space—so there is also no unique intuition of time that applies to all mankind. Not only have different civilizations assigned different degrees of significance to the temporal mode of existence, but the mode itself has been regarded in different ways.

Nevertheless, even if their lives are not dominated by time in the way that ours have come to be in recent centuries, all peoples, however primitive, have some idea of time and some method of reckoning it, usually based on astronomical observations. The Australian aborigine, for example, has great difficulty in telling the time of day by a clock, and even if he learns to read off the position of the hands as a memory-exercise, he does not relate it to the time of day. On the other hand, he is capable of fixing the time for a proposed action by placing a stone in the fork of a tree, or some such place, so that the sun will strike it at the agreed time. His difficulties with the clock are cultural.

Time and the origins of scientific thought

The late emergence of the concept of time as a major influence in the sciences did not result from any deficiencies of temporal awareness by our predecessors. Instead, it can be explained as an inevitable consequence of the origins of rational thought itself. In his classic work on *Primitive Man as Philosopher* Paul Radin[4] argues that among primitive men there are two different types of temperament: the man of action who is oriented towards external objects, interested primarily in practical results and comparatively indifferent to the stirrings of his inner self, and the thinker—a much rarer type—who is impelled to analyse and 'explain' his subjective states. The former, in so far as he considers explanations at all, inclines to those that stress the purely mechanical relations between events. His mental rhythm is characterized by a demand for endless repetition of the same event or events, and change for him means essentially some abrupt transformation. The thinker, on the other hand, finds purely mechanical explanations inadequate. But, although he seeks a description in terms of a gradual development from one to many, simple to complex, cause to effect, he is perplexed by the continually shifting forms of external objects. Before he can deal with them systematically he must give them some permanence of form. In other words, the world must be made static.

Belief that ultimate reality is timeless is deeply rooted in human thinking, and rational investigation of the world originated in the search for the permanent factors that lie behind the ever-changing pattern of events. Indeed, language itself inevitably introduced an element of permanence into a vanishing world. For although speech is based on sound, and sound is transitory, the conventionalized sound-symbols of language transcended time. At the level of oral language, however, permanence depended solely on memory. To obtain a greater degree of permanence, the time symbols of oral speech had to be converted into the space symbols of written speech. The earliest written records were simply pictorial representations of natural objects, such as birds and animals. The next step was the ideograph, by means of which thoughts were represented symbolically by pictures of visual objects. The crucial stage in the evolution of writing occurred when ideographs became phonograms, that is representations of things that are heard. This conversion of sound-symbols in time to visual symbols in space was the greatest single step in the quest for permanence.

The development of rational thought and the origins of the scientific world-view can be regarded as an extension of the quest for permanence already revealed in the formation of language and the invention of writing. Generally speaking, for the Greek thinkers who were responsible for taking the first steps towards modern science, the concept of time was not of primary importance. Even Heraclitus, who regarded change as the very essence of reality, adopted a concept of transmutation involving simultaneous processes of creation and decay that maintained a permanent and not a progressive order of the cosmos. Similarly, Anaximander, who is regarded as the author of the first scientific cosmogony and of the first theory of organic evolution, appears to have believed in the continual generation and destruction of worlds as a permanent process without beginning or end. Indeed, his conception of the cosmic process was not evolutionary at all. Instead, it involved the cyclic alternation of opposites. According to the commentator Simplicius, Anaximander said, 'The source from which existing things derive their existence is also that to which they return at their destruction, according to necessity; for they give justice and make reparation to one another for their injustice according to the ordinance of time.'[5] Thus the idea of time was associated with the idea of 'justice' or balance rather than progress, and with the theory of primary opposites rather than a single linear variable. The whole conception was no doubt suggested by the cycle of the seasons, with its alternating conflict of the hot and the cold, the wet and the dry. Each of these advances in 'unjust' aggression at the expense of its opposite, and then pays the penalty, retreating before the counter-attack of the latter, the object of the whole cycle being to maintain the balance of justice.

The concept of nature as a process of continual strife of opposite powers subject to the ordinance of time was submitted to penetrating criticism by Parmenides, the first philosopher to formulate logical proofs rather than dogmatic aphorisms. He was obsessed with the idea that change involves contradiction, and so is logically incompatible with existence. He therefore regarded the idea of time as irrational or contrary to reason. Parmenides had a considerable influence on Plato, in whose cosmological dialogue, the *Timaeus*, space and time were treated very differently; space existed in its own right as a given frame (the 'receptacle') for the operations of reason (the 'demiurge') in producing the visible order of things, whereas time was merely a feature of that order based on an archetype ('eternity') of which it was the 'moving image'. The archetype was an ideal (or, as we would now say, 'theoretical') realm of timeless existence, exemplified by perfect geometrical figures, geometry having become by this stage in Greek thought a logical science.

In the physical universe, where Plato recognized that the concept of time could not avoided, he maintained that cyclic processes must predominate. The essentially cyclical nature of change was also stressed by Plato's more empirically minded pupil Aristotle. Unlike Plato, Aristotle did not believe in the existence of a transcendental timeless world of ideal forms, and he thought that the proper object of scientific investigation was the world revealed to us by the senses. He rejected altogether the idea of Plato that the order of nature was mathematical in character. He defined mathematics as the science of changeless things, and physics as the science of things that do change. Consequently, in Aristotle's view physics could not be applied mathematics. Although he regarded motion and change as the essential characteristics of the physical world, his concept of motion was very different from ours. His primary concern was with the states before and after motion rather than with the dynamic process itself, and his outlook was teleological. He believed that there were final causes in nature and that even the behaviour of inanimate matter was determined by goal-seeking tendencies. Every body had its natural place in the universe, depending on its composition, and if displaced therefrom had a potential tendency to return to it. Thus some bodies had a natural tendency to fall, others to rise. The heaviness of earth and water and the lightness of air and fire were manifestations of these potential tendencies. Similarly, the essence of living creatures was not the process of growth but the form of the fully developed organism. Form and place were therefore more fundamental than time. Aristotle's teleological outlook, with its emphasis on end-conditions, necessarily meant that for him time played a far more subsidiary role than in modern science, where we think in terms of invariable sequences rather than essences, and investigate laws of nature that determine the development of systems in time from given initial conditions.

That the static form rather than the dynamic process was the fundamental concept in Aristotle's philosophy of science no less than Plato's, despite the empirical tendency of the one and the abstract tendency of the other, was in accord with the general spirit of Greek scientific research in practice. For Greek scientists mainly concentrated on the study of static relations between phenomena. In acoustics they discovered the correspondence between the pitch of a note and the length of a cord, but they did not proceed to calculate pitch on the basis of a theory of vibrations. In optics they studied only the geometry of light-paths and never arrived at Fermet's principle of least time. In mechanics, despite the attempt of Aristotle to analyse the motion of falling bodies and projectiles, their only positive achievements were confined to statics and hydrostatics, as investigated by Archimedes.

The evolution of the concept of mathematical time

It is against this historical background that the powerful concept of mathematical time originated and developed, as discussed in the first essay of this book. It can be traced back to the same era as the invention of the mechanical clock—the early fourteenth century. Even if Greek thought had been less dominated by the example of geometry and had been more influenced by dynamic concepts, the development of the modern metrical concept of time and its practical application would still have been greatly impeded by the lack of reliable mechanical clocks for its accurate measurement. The only mechanical time-recorders in antiquity were water-clocks, and until the fourteenth century the most reliable way to tell the time was by means of a sundial. Although the accuracy of the new mechanical clocks was low by modern standards, they made a tremendous impression on European society, particularly because of the elaborate astronomical and other gadgetry associated with them. As a historian of mediaeval technology has remarked, 'No European community felt able to hold up its head unless in its midst the planets wheeled in cycles and epicycles, while angels trumpeted, cocks crew, and apostles, kings and prophets marched and counter-marched at the booming of the hours.'[6] This reaction indicated a shift in the values of European society, the first sign that the concept of mathematically divided time was to become the new medium of human existence. In the same century, the rigid distinction that Aristotle had made between mathematics and physics was abandoned, and the first steps were taken towards the mathematical analysis of continuous change and the idea of time as a mathematical variable.

Nevertheless, the development of the idea of mathematical time and the growth of its influence on society were far from rigid. Traditional cyclic ideas of astral influences still prevailed, and since even the Church endowed each date in the year with some symbolic significance the ecclesiastical calendar tended to reinforce the conviction that time was uneven in quality. In other words, for most men 'magical' time had not been superseded by homogeneous mathematical time. Moreover, throughout the whole a mediaeval period, there was a conflict between the cyclic and linear concepts of time. Scientists and scholars influenced by astronomy and astrology emphasized the cyclic concept, whereas the linear concept was fostered by the mercantile class, as a money economy came into being. So long as power was concentrated in the ownership of land, time was associated with the unchanging cycle of the soil. With the circulation of money, the emphasis came to be placed on mobility, and men began to believe that 'time is money' and should therefore be used economically. Thus, time came to be associated with the ideas of work and of linear progress, and this view was reinforced by religious tendencies in the Renaissance.

The invention of a successful pendulum clock in the seventeenth century had a tremendous influence on the concept of time, for at last mankind was provided with an 'accurate' timekeeper that could tick away continuously for years on end. This must have greatly strengthened belief in the homogeneity and continuity of universal time. The idea of mathematical time on which Newton based his famous laws of motion and gravitation possessed these properties, being a continuous sequence of moments analogous to a geometrical line. He thought of time as being not only universal but also 'absolute', that is to say, existing in its own right independently of all particular events and processes. Although by the end of the nineteenth century it was recognized by some physicists that this idea was unnecessary, and that it was sufficient to regard mathematical time as an abstract concept implicitly defined by the laws of motion, everyday life, at least in the countries influenced by European and American technical developments, had come to be dominated by the conviction that absolute time was a concrete reality. This was due to the increasing subordination of men's lives to the tyranny of the clock. Even the division of the earth's surface, in 1885, into separate time-zones did little to undermine this widespread and almost instinctive belief.

As the first essay explains, the theory of relativity has undermined the idea of universal absolute time and has replaced it by the idea that different observers in relative motion will, in general, assign different times to the same event. As a result, a moving clock will seem to run slow compared with an identical clock at rest with respect to the observer. The theory has had an important impact on the concept of causality, since it implies that no causal influence can be transmitted with a speed exceeding that of light. However, the theory's alleged implication that the passsage of time is a purely subjective feature of human consciousness, and that there is no physically significant time-scale of the universe, has been shown to be groundless. Indeed, the theory has led to time-measurement becoming even more significant than before in physics. Nevertheless, these sophisticated considerations have so far had little effect upon the average man's idea of time, which still tends to be more in line with the point of view that prevailed in the nineteenth century.

Time in astronomy and astrophysics

In the second essay, we pass on to consider how discoveries in astronomy and astrophysics have influenced our ideas about time. Professor Piotrowski considers first the consequences that have been drawn in recent times from the familiar observation that the night sky is dark, and he shows that this is intimately connected with cosmological problems that were not resolved until the unexpected discovery, by the American astronomer E. P. Hubble in 1929, that the universe as a whole is expanding. Since the galaxies, those great stellar systems which we believe to be the major 'building blocks' of the universe, are receding from each other, and consequently from us, the light we receive from them is much dimmer than it would otherwise be, and this accounts for the darkness of the night sky. The expansion of the universe can itself be regarded as an indicator of time's arrow, for if the latter were reversed expansion would be replaced by contraction and the night sky would be bright. The darkness of the night sky thus provides an important link between time and the universe. This is particularly significant since, with one minor exception, the basic laws of physics, such as the law of gravitation, are unchanged when the direction of time is reversed; consequently, they provide no indication of time's arrow.

The way in which astronomy most directly affects everday life is in the measurement of time. Throughout history, the ultimate standard of time has been derived from astronomical observations. In due course this led to the hour, minute, and second being defined as fractions of the period of one rotation of the earth on its axis, as determined by the careful study of the apparent daily rotation of the celestial sphere. In particular, what is called 'sidereal time' is based on the times of transits of stars across the meridian. (The meridian is the projection on the sky of the circle of longitude through the place on the earth's surface where the measurements are made.) The interval between successive transits of the same star, or group of stars, is the sidereal day. The precise standardization of time measurement in this way dates from the foundation of the Royal Observatory at Greenwich in 1675. The need for accurate time in those days was primarily felt by navigators, for without an accurate clock that could keep Greenwich time it was impossible to determine the longitude of a ship at sea, and it often happened that a ship was hundreds of miles off course. The successful transportation of Greenwich time to any place on the globe was eventually effected by the perfection of the marine chronometer, about 1760, by John Harrison.

As Professor Piotrowski points out, however, more important for science and culture than the measurement of seconds or days is the determination of very long intervals of time. One of the greatest obstacles that scientists had to contend with, particularly in the eighteenth and nineteenth centuries, was the widespread inherited conviction that the range of past time was severely limited. For example, Archbishop Ussher in 1658 calculated that God created the world on Sunday, 23 October 4004 B.C. It is remarkable what a strait-jacket this Bible-based chronology was for scientists studying the nature of fossils. Nevertheless, during the eighteenth century, scientists and others began to think of the past in terms of millions rather than of thousands of years. In the following century this point of view, and with it the general idea of time as linear advancement, ultimately prevailed through the influence of the biological evolutionists, but the climate of thought that made it possible to contemplate the hundreds of millions of years required for the operation of natural selection to account for present and past species was prepared primarily by the geologists.

The most serious objection to their demands on past time came, however, not from the theologians but from the physicists. Estimates made by them of the age of the earth, based on thermodynamic considerations concerning the earth's rate of cooling, feel far short of the demands of the geologists. The resolution of this difficulty was possible only after the discovery of radioactivity and the investigation of nuclear transformations by Rutherford, and others, early in this century. It is now known that there is a sufficient supply of radioactive elements in the crustal rocks to make the net loss of heat from them extremely small. As a result, the age of the earth is now estimated to be between four and five thousand million years. Similarly, since the investigations of Bethe and Weizsaecker in the late 1930s, it is generally accepted that the sun's heat is maintained by thermonuclear processes that can continue steadily for thousands of millions of years.

This discovery of the systematic recession of the galaxies, which has been generally interpreted as an expansion of the whole universe, has led to estimates for the time that has elapsed since the initial stage when the whole cosmic evolutionary process began. These estimates depend on the particular assumptions made concerning how the rate of expansion in the past compared with its rate today—for example, whether the expansion has been uniform in time or is slowing down. All such estimates, however, give the same order of magnitude for the age of the universe, roughly ten thousand million years. In the present state of knowledge, this is the longest stretch of past time over which we can extend the laws of nature as we know them.

In the final paragraph of his essay, Professor Piotrowski sounds a somewhat pessimistic note concerning the impact of recent advances in astronomy and astrophysics on mankind generally. It is true that the highly sophisticated technical details of this research have made little impression except on scientists, but there is a fairly widespread educated public that shows great interest in the main results of modern astronomy and cosmology. This is evident from the success of many popular radio and television programmes on these subjects broadcast in recent years. Indeed, of all branches of science, except those concerning health and medicine, there is no doubt that astronomy has the greatest appeal to the average intelligent man. Just as the idea of biological evolution had a tremendous impact on men of the last century, so the concept of cosmic evolution, with its implications as regards the time-scale of the universe, has been a major influence on modern man's world-view.

Time, biology and human evolution

When we turn to biology and human evolution we are confronted by a very different set of time-scales. Enzymatic reactions occur in times of the order of 10^{-4} seconds, general physiological reactions occur in times of the order of seconds, cell divisions take several hours, the life of man is of the order of ten to a hundred years, whereas the time-scale of evolutionary change must be reckoned in terms of millions of years. Dr Vester, in his essay on 'Time and biology', is concerned with the different time-scales associated with the different phases of human evolution, particularly the later phases. Life has existed on earth for more than 3,000 million years, but man has existed for a time of the order of only 1 to 2 million years. Ninety-nine per cent of this time he has spent as a hunter and collector of food. Only in the last 10,000 years has he become an agriculturalist and developed civilization. Dr Vester stresses the point that man's present genetic equipment must, therefore, have been mainly influenced by natural selection in the period when he was a hunter and food-collector, that is to say, the period before civilization. With the introduction of agriculture there was a sudden change in man's relation to his environment. The myth of the 'fall of man' refers to this separation of the human race from the rest of nature, as the result of the acquisition by man of knowldge that he did not possess previously. Another relatively sudden major change has occurred in recent centuries, with the application of scientific and technological knowledge to the transformation of the environment. Since these abrupt changes have occurred while man retains the same genetic equipment he had a million or more years ago, it is not surprising that he has developed what Dr Vester calls 'adaptation syndromes'.

At the present time, one of the most significant features of our social evolution is the concentration of more and more people in cities and urban areas. Whereas in the year 1800 only about one person in forty in the world was a town-dweller, today the proportion is about one person in three, and this proportion is growing rapidly. In the last century civilization did not span the whole world in the way it does today, when what happens in one part may affect any other part. Moreover, the higher the technological development, the greater the demands made on the environment. To take the most striking example, in the last quarter of a century the population of the United States has increased by a half, but the burden on the environment of each inhabitant has increased seven-fold. Since in a world of limited resources this rate of consumption cannot continue indefinitely, the paramount need is to develop the technology of re-cycling. Consequently Dr Vester lays particular stress on cybernetic technology involving negative feedback, which has a stabilizing effect (an excessive value of the variable to be controlled being reduced by the regulator, whereas too low a value is increased).

As far as concepts of time are concerned, two complementary ideas emerge from Dr Vester's paper: evolution and cycles. He is critical of the modern tendency to concentrate on the idea of time as linear advancement, that is, of time as an arrow, because he believes that it prevents us from paying due regard to cyclic processes. Instead of thinking of events only in terms of linear cause-and-effect relations, which in his terminology are only 'short arrows', we should consider longer time intervals in which events may constitute a feedback cycle. However, although he urges us to pay due recognition to cycle processes, he is careful to point out that we must not overlook linear evolutionary processes. What is required is neither the one nor the other but a genuine combination of the two. For, if we are to avoid catastrophe, we must engage in long-term planning for fifty, or even a hundred, years ahead. A similar need arose in the past when there was another abrupt major change in man's way of life. For, whereas the primitive hunter needed to think at most only a day or two ahead, when man invented agriculture it became necessary to think up to a year ahead, although agriculture itself is a cyclic process. To cope with the present crisis of our civilization we must, according to Dr Vester, not only be prepared to think and plan fifty to a hundred years in advance, but our way of thinking must be cybernetic. Our cybernetic thinking should not, however, be based on rigid control engineering, since living processes take place in an assembly of interacting open systems with feedback control cycles, and it is this interlocking self-regulation that we should seek to emulate in our evolutionary feedback strategy for the future of civilization.

Time and human psychology

The effect on human behaviour of time and of man's awareness of time are discussed in the essay by the distinguished French psychologist Paul Fraisse.

Animals as well as man are affected by the daily and annual cycles that are imposed upon us because we live on the planet earth. All our activity is subject to a 24-hour rhythm. For example, it is well known that our internal temperature is normally at its highest in the evening, and lowest early in the morning. Indeed, there is apparently no organ in the human body that does not display a similar rhythmicity. As Professor Fraisse remarks, our organism is a biological clock with its own regulator, but capable of being set to a different time by an external synchronizer. This clock has been tested by placing subjects in underground caves where they are affected by few external stimuli. Although their estimation of time is often badly affected, the number of activity-sleep cycles is usually maintained with remarkable accuracy. Further evidence for our possession of a biological clock is revealed by the desynchronization that occurs when we make a long east-west or west-east flight by air, or when we are subjected to an abnormal time-schedule. Such phenomena are most readily interpreted on the assumption that an internal clock is involved which is out of phase with the time of the environment.

Many of our activities can be considered as responses, not to cycles, but to a regular succession of signals. This is observed in the training of young children, as they become adapted to routine successions of movements—for example, when being dressed. In general, we learn various temporal sequences so as to avoid being taken by surprise by foreseeable events. This temporal regulation affects us at different levels, for it not only enables us to forestall fatigue, but also, through our sensori-motor activities, to anticipate external stimuli.

We have already seen that man's relation to time is quite different from that of animals, because he is not only subject to time, but is also conscious of time. Professor Fraisse discusses four aspects of our temporal awareness, beginning with the psychological present. This differs sharply from the mathematically precise point-like instant, separating past from future, associated with the continuous time variable t used in theoretical physics, since it has temporal extension. As he points out, were this not the case, we would be unable to perceive a melody, for all we would be aware of would be an unrelated succession of notes. More generally, we can assert that our direct perception of change depends on the simultaneous presence, in our awareness, of events in distinct phases of presentation. Fraisse considers that the duration of the psychological present may last up to five seconds. He also considers our estimation of duration, and why it so often bears little relation to the corresponding time interval measured by a clock. The reason is that our powers of temporal estimation are influenced by the tempo of our attention, and many factors can affect this. Often we would do better to rely on our biological clock than to invoke our sense of judgement. The 'head clock', as it is often called, works with considerable success under hypnosis and even without hypnosis it can often function precisely over fairly long periods. In a classic experiment Macleod and Roff[7] found that two subjects shut up in a sound-proof room, for 48 hours and 86 hours respectively, estimated the time with such accuracy that their respective errors amounted to less than 1 per cent.

The other aspects of our temporal awareness discussed by Professor Fraisse are what he calls 'the temporal horizon' and the genetic development of the concept of time. The former bears directly on the topics discussed by some of the other contributors to this book. For man's temporal horizon is determined by his knowledge of the past and his ability to predict the future. It depends on his age, and particularly on the nature of the culture to which he belongs—a pervasive influence on the way in which time is regarded. The genetic development of the idea of time also bears on some of the topics discussed elsewhere in this book. For the development of the growing child's understanding of time has some similarity to the increasing sophistication of mankind's temporal concepts. In an appendix, Professor Fraisse discusses topics, such as teaching the concept of time to children, the way in which work is organized, and man's attitude towards the future, which all depend on the cultural level of individuals and of the community as a whole.

Time and the future

It is a plausible hypothesis that the original source of man's idea of time was an accumulation of sensations that produced an internal perspective directed towards the future. For the great development of the prefrontal lobes of the brain of *homo sapiens* was probably closely associated with his growing power of adjustment to future events. It is possible that Neanderthal man may have had some rudimentary concern for the future, since it appears that he buried his dead, but the emergence of modern man has been correlated with a strongly increased tendency to look forward. Evidence for this conclusion is provided by the sudden development of tools that appear to have been used to make a wide variety of other tools, such as barbed harpoons and fish hooks, for future use.

    As is pointed by Dr McHale in his essay on 'Time and the future sense', although in ancient civilizations the myths that were ritually re-enacted periodically were ostensibly concerned with the past—for example, the origin of things—the object of their re-enactment was to serve the future. In our own age, the ascendancy of science over religion is partly due to the widespread belief in its greater predictive power. Our concept of an unfolding linear future is, as has already been stressed, a comparatively recent development. Closely connected with it is the idea of social progress and the improvement of the human condition. In the past, most societies adopted a cyclical model of future change, but Dr McHale claims that our current views derive from the Greeks and certain mid-eastern religions. Although he refers to the Greeks simply because of their development of the idea of rational inquiry, which is a basic feature of the modern scientific world-view, it should be mentioned that in his posthumous book *The Idea of Progress in Classical Antiquity* the eminent classicist, medical historian, and philosopher Ludwig Edelstein argued that Epicureanism, Scepticism and Stoicism, the three dominant philosophical schools in the later Hellenistic period, all embraced progressivism in some form or other.[8]

Coming nearer to our own times, Dr McHale refers to the influence of the Renaissance and the Reformation on the idea of material progress, but strictly speaking the former is meant to include the Scientific Renaissance—often called the Scientific Revolution—of the seventeenth century as well as the earlier Humanistic Renaissance. In the eighteenth century both the Enlightenment and the beginnings of the Industrial Revolution generated greater expectations of a better future. These expectations culminated in nineteenth-century optimism. Dr McHale draws attention to the present paradoxical situation in which the developed world has become disenchanted with the idea of material progress, whereas the lesser developed countries have vigorously embraced it. This disenchantment in the developed world with the idea of progress and the current emphasis on the 'limits of growth' are due to the acceleration of change that we have experienced in the last hundred years. A greater realization of the 'side effects', or unintended consequences, of rapid change has produced disillusionment. Optimism concerning the future has been replaced by anxiety. This has led to much technological and social forecasting and planning, but as McHale points out, there are difficulties in going from the world of technology, in which physical laws play a major role, to the sociological world, in which change is far less deterministic. He criticizes many recent attempts at forecasting events in the latter world, because it is often assumed that social processes are more analogous to physical processes than in fact they are. In particular, he considers that there has been too little awareness of the effects of changing concepts of time on social groups and individuals.

McHale concludes by drawing attention to the great increase in recent years in the number and frequency of conferences, etc. devoted to the formal study of the future. This flowering of interest, he suggests, reflects a major change in temporal perspective. But to what degree, he asks, can we justify the sacrifice of present time to the more speculative requirements of the future?

Time and information science

Some fifty years ago Alfred Korzybski introduced a theory of organic and human behaviour based on what he called 'binding' ability, all organisms being binding agents, the term 'binding' signifying 'tying together', 'using', 'transforming', etc. Whereas the binding ability of plants is purely chemical, since they are relatively immobile in space and incapable of group behaviour and communication, and animals are both chemical-binding and space-binding, because they are capable of movement and simple group behaviour but incapable of communication from gene-ration to generation, man is also time-binding, since he can transmit information through time. Man's time-binding ability results from his use of language and symbolic forms, for these make the communication of information possible. Our predictive capacity and planning ability have been greatly extended this century by the application of cybernetics, information theory, general systems theory and general semantics to science and technology. As a result, man is now beginning to influence his future and direct his own evolution more effectively than at any time in the past.

Communication, like other processes, is a time-dependent pheno-menon. In his essay on time in information science, Dr Neelameghan draws attention to the close parallel between the transmission of knowledge and of disease, so that the communication process can be regarded as analogous to an epidemic. He points out that only one of the models so far proposed for a mathematical theory of the communication process involves time explicitly, and this model is essentially a diffusion process. Time is involved because there are time lapses between in-novation and the first individual decision to make use of it, in the rate at which it is then adopted by others, measured as the number of adopters in a given period of time, and the degree to which an individual is relatively earlier than others in his social system to adopt new ideas.

The acceleration in the rate at which new knowledge has been put to use, particularly following the great expansion of scientific research during and since the Second World War, has also led to an appreciable acceleration in the rate at which knowledge has become obsolete. Neelameghan emphasizes the dangers of too much emphasis on speed of communication leading to a flood of information of unequal value confronting the information-seeker, and he quotes an American es-timate that, at the rate at which it is now growing, the amount of knowledge in the world fifty years hence will be thirty-two times as much as it is now. Although he does not explain how this calculation is made (that knowledge 'doubles' every ten years), it is difficult to dissent from his conclusion that the amount of information at man's disposal is likely to present increasingly serious problems for the decision-makers. Moreover, the increase in the speed of transmission of information exerts an additional pressure. Dr Neelameghan's paper concludes with an assess-ment of the effects that this 'information explosion' is likely to have on the individual. His picture of the future complements that drawn by Dr McHale.

The concept of time in rural societies

In his paper on the concept of time in peasant society, Dr Abou-Zeid points out that for the peasant it has been customary for life to be organized according to the seasons rather than by hours or smaller units of time. Consequently, the pace of everyday life has been much slower than in industrialized society. Nowadays, however, he is usually confronted by a diversity of systems of time-reckoning, for besides retaining one or more traditional calendars, most peasant societies have also adopted the Western calendar. The former are generally used for regulating traditional social activities and the latter for regulating the more modern aspects of life, especially those relating to government. In the author's own country, Egypt, three different calendars are used by the peasants: the Coptic, the Islamic and the Western. The first is a solar calendar that is probably based on the ancient Egyptian system of time-reckoning and is used for agricultural purposes. The second is lunar and is used for religious ceremonies and festivals. The Western calendar was formerly little used by the peasants, but recently the younger peasants, who have been to the State schools, are tending to use it. As Dr Abou-Zeid remarks, 'this represents an important change in the traditional system of time-reckoning'.

In peasant society time is a primarily a system of events that concern the community as a whole rather than the individual. Nevertheless, it is a mistake to assume that peasants live only in the present moment. Instead, they regard the past with a sense of nostalgic admiration; the days that have passed are the best, for we know all about them. On the other hand, the future is illusory, for it is characterized by uncertainty; it does not belong to men but only to God. Nevertheless, despite the uncertainties that beset the individual and occasionally the community, the future is generally regarded as being more or less a repetition of the past. This cyclical image of time, which has been almost universal, raises doubts about the ability of the peasant to plan for the future. Instead, planning for a future that is radically different from the present falls beyond the scope of the old-style peasant. Because of the introduction of compulsory free education in Egypt, however, the young villagers are tending not only to use mainly the Western calendar but are correspondingly more receptive to the idea of planning for the future. The other important influence that is changing the peasant's attitude to time is the radio. Those who listen regularly to particular programmes begin to respect punctuality and precise timing.

Professor Kiray in her paper gives a careful analysis of the way in which rural societies are affected by the two fundamental aspects of time—sequence and duration. In these societies, in each cycle, the successive main points of change are marked by 'rites of passage'. The repetitive patterns of human life and the natural world provide measures of duration. Although reckoning by days, months and years occurs in all rural societies, these units are not usually regarded as forming a related set of measurements. Whereas the day, month and year are based on natural phenomena, the week (which varies in length from one society to another) is an entirely social construct, and one day in it is usually allocated for religious activities. Consequently, whereas the basis of the week is often economic, that of the rest-day is religious.

In 1962 Professor Kiray made a study of the concept of time in various rapidly changing rural communities in northern Turkey, where farming had been affected by the introduction of cash cropping and an efficient road transport system. She found that, as one passes from more isolated to less isolated communities, or as the community becomes more developed, so international units of time are used more, roughly in proportion to the degree of impact of modern technology, and particularly modern transport. Moreover, literacy influences attitudes to the past and future as well as the present. In this context there is a link with Dr Neelameghan's essay, for she points out that the permanent nature of written records makes a great difference to the accumulation and storage of knowledge, and so creates the possibility of rapid change.

Time in rapidly changing societies

The questioning of values is exemplified by the essay of Josefina Mena Abraham. Rather than time in the sense in which it is used by other writers in this volume, she applies herself to the idea of change, that is to say, to what it is that happens in societies in the temporal framework. Her essay differs from the others in another respect, namely that she employs the methods of a particular system of political philosophy (her own adaptation of Marxist principles) to analyse problems of change in several kinds of co-existent society. She begins by considering the nature of perception, taking sensory perception as a starting point, moving on to the perception by the individual of his environment and culture, and then developing arguments about several kinds of change in societies: spontaneous change, change imposed by alien intervention, change considered as a reaction against alien intervention. Readers to whom this type of dialectical analysis is unfamiliar may find this essay hard to follow. Those who habitually think in this way will find it easier to relate the essay, including as it does illustrations from Latin America, India and other areas under stress, to the conditions described, from quite a different standpoint, by Professor Kiray. Ms Abraham does not, however, take examples from Africa, where the stresses connected with time and change differ in many respects.

Dr N'Sougan Agblemagnon is concerned with the problem of time in Africa. He argues that the African world is a world under stress, subject to conflicting forces and continually changing. The time of the individual is experienced in a social context. This means that, however much he strives to be modern, he is still subject to traditional influences. Consequently, he is affected by a duality of time systems. Traditional African time, as exemplified by the futurology of the soothsayer, is associated with dreams and may be regarded as 'mythical time', in contrast to Western scientific time, which is associated with the rationalistic, statistical futurology of the economist. Whereas African time helped to integrate the individual with the community, Dr Agblemagnon maintains that Western scientifically based time isolates him from the social setting. Although this may be true in Africa, it is certainly not the case in Europe and North America, where men's lives are controlled by time based on the time service relayed from national observatories. In these societies, scientifically controlled time is a major influence towards integrating the individual with the community. Without it, much that characterizes Western civilization today, such as public transport, punctuality for appointments, etc. would be made much more difficult.

The increasing influence of time on mankind

The rapid growth of science and technology and the influence they have come to exert in almost all parts of the world have led to the widespread adoption of the linear progressive metrical concept of time, which conflicts with the traditional cyclic concept that formerly helped to keep a society stable. Moreover, in modern industrial society time is no longer a concept of secondary importance. It is the dominant feature, not only of our world-view, but of our social organization and our way of life. As Lewis Mumford has said, 'The clock, not the steam-engine, is the key-machine of the modern industrial age.'[9] By its very nature, it has helped to dissociate time from its dependence on particular human events and to foster belief in a world of mathematically measurable sequences—the world of science. With the rapid advance of science and technology, all our activities have tended to become controlled more and more by time. The popularization of timekeeping that followed the mass-production of cheap watches in the nineteenth century has accentuated the tendency for even the most basic functions of living to be regulated chronometrically: one eats and sleeps, not just when one is hungry or tired, but when prompted by the clock. Also, instead of cyclic repetition being regarded as the characteristic feature of time, increasing emphasis has come to be laid on the distinctions between past, present and future. In particular, more and more attention is being paid to prediction and forecasting. With the general realization that the world population is increasing ever more rapidly, although we inhabit a world of limited resources, the future of mankind has become a question of urgent concern. To understand the nature of modern industrial society and the problems that beset it, we must take into account the effect of human thought and behaviour of the modern scientifically based concept of time.

NOTES

1. E. Cassirer, *An Essay on Man*, p. 51, New Haven, Conn., Yale University Press, 1944.
2. W. Koehler, *The Mentality of Apes*, p. 230 et seq., Harmondsworth, Penguin Books, 1957.
3. B. L. Whorf, 'An American Indian Model of the Universe', in *Language, Thought and Reality*, p. 57–64, Cambridge, Mass., MIT Press, 1956.
4. P. Radin, *Primitive Man as Philosopher*, p. 230 et seq., New York, Dover Publications, 1957.
5. Simplicius, *Phys.*, 24, 17.
6. Lynn White, *Medieval Technology and Social Change*, p. 120, Oxford, Clarendon Press, 1962.
7. R. B. MacLeod and M. T. Roff, 'An Experiment in Temporal Disorientation', *Acta psychol.*, Vol. 1, 1936, p. 381–423.
8. L. Edelstein, *The Idea of Progress in Classical Antiquity*, Baltimore, Md., Johns Hopkins Press, 1967.
9. Lewis Mumford, *Technics and Civilization*, p. 14, London, Routledge & Kegan Paul, 1934.

MATHEMATICAL TIME
AND ITS ROLE
IN THE DEVELOPMENT
OF THE SCIENTIFIC
WORLD-VIEW

Gerald J. Whitrow

The civilization of Western Europe has produced in recent centuries two unique contributions to thought that have transformed man's conception of the universe and his way of life: the scientific world-view and the idea of time as linear advancement, associated, in practice, with the regulation of our lives by the clock and, in theory, with the dominant role played by the continuous variable t, denoting mathematical time. It is the object of this essay to survey the evolution of the concept of mathematical time and the way in which it has influenced, and been influenced by, the development of our general conception of the universe.

The origin of mechanical clocks and mathematical time

The representation of time as a fundamental mathematical parameter was not a feature of ancient science. For the Greeks, as for other civilizations before our own, the essentially cyclical nature of change was a basic assumption that characterized their attempts to understand the universe. This led to time being regarded as a concept of secondary significance and the temporal aspects of phenomena as subordinate to the permanent. Scientific investigation was the quest for 'essences', that is for the underlying changeless qualities, or essential nature, of things. Unlike the Greeks, we tend to think in terms of sequences rather than essences, and we investigate the development of systems in time from given initial conditions. But, even if Greek thought had been more influenced than it was by dynamic concepts, the development of the modern metrical idea of time would still have been greatly impeded by the lack of reliable clocks for its accurate measurement.

Although there is no evidence of any direct link, it is presumably no coincidence that the origin of the modern mathematical concept of time can be traced to the same era as that of the invention of the mechanical clock, namely the early fourteenth century. This soon led to the modern system of time-reckoning in which day and night together are divided into twenty-four equal hours. Previous civilizations did not have a uniform scale of hours. Instead, the periods of light and darkness were each divided into an equal number of so-called 'temporal hours'.

Consequently, the length of an hour varied according to the time of year and, except at equinoxes, a daylight hour was not equal to a nocturnal hour. Only in theoretical works of Hellenistic astronomy were 'equinoctial hours' of equal length introduced. They were the same as the temporal hours at the date of the spring equinox.

Until the middle of the seventeenth century, mechanical clocks had only one hand, and the dial was divided only into hours and quarters. The division of the hour into sixty minutes, and the minutes into sixty seconds, was used in 1345 to express the duration of a lunar eclipse, but this was the result of a computation.[1] Nevertheless, despite the practical difficulties of precise time-measurement, the abstract framework of mathematically divided time gradually became the new medium of daily existence, so that, as Lewis Mumford has stressed, 'Eternity ceased gradually to serve as the measurement and focus of human actions.'[2] Mathematical time was also destined to become the medium for man's conception of the universe, but this only came about very slowly. For, despite the influence of Christianity, with its emphasis on the Incarnation as a unique event not subject to repetition, the theory of cycles and of astral influences was accepted by most Christian thinkers until the seventeenth century. In their world-view, time was not conceived as a continuous mathematical parameter, but was split up into separate seasons, divisions of the zodiac, and so on, each exerting its specific influence.

The origin of the modern mathematical concept of time is associated with that of the general mathematical notion of the variable, resulting from the detailed criticism of Aristotelian philosophy in the early fourteenth century. Aristotle had made a rigid distinction between mathematics and physics, the former being concerned with 'things which do not involve motion' and the latter with things that do. Moreover, since in terrestrial motions, unlike celestial, there appeared to be no general uniformity, physical motion was regarded by Aristotelians, not as a 'quantity', but as a 'quality' that does not increase or decrease through the joining together of parts. Duns Scotus, who died in 1308, was among the first who broke with this tradition and considered the general problem of the variability of qualities.

The need to abandon Aristotle's dogma of the immutability of substantial forms arose in connection with the concept of intensity. For example, if we increase or diminish its intensity, a ray of light will become more or less luminous without there being any change in its nature, which is simply light. Duns Scotus and his followers at Merton College, Oxford, held that increase in intensity occurs by addition, like water added to water. The original analysis was verbal and prolix, but it merged into the mathematical problem of describing the various possible modes of spatial or temporal variation of intensity.

The pioneer figure in the mathematical development of the idea of the variable was Thomas of Bradwardine, whose *Tractatus de Proportionibus* (Treatise on Proportions), written in 1328, is noteworthy for introducing into mathematics functions other than those associated with simple proportionality. Later in the century Nicole Oresme, in France, made a fundamental advance when he discussed continuous change with the aid of a two-dimensional diagram. A finite horizontal line represented the extension of a given form. This line, called longitude, was divided into equal divisions called degrees. An intension, or rate of change, by which a form acquired a quality or property was represented by a vertical line, called latitude, drawn on a uniform scale through the corresponding longitude. When all the vertical lines had been plotted, a line drawn through their summits yielded a geometrical figure which Oresme called the linear configuration of the quality under consideration. This line of summits foreshadowed the idea of the curve representing a function in analytical geometry, but for Oresme the function was represented by the area under the curve, and the ordinates or vertical lines represented rates of increase of the function rather than successive values of it.

Like his predecessors in Oxford, Oresme realized that the longitude could be taken to represent time and latitude velocity, the area under the curve (with appropriate end conditions) then representing the distance traversed in a given time. Despite his mathematical deficiencies, notably his lack of algebraic ideas and symbolism, Oresme's achievement marks an important transitional phase in the evolution of theoretical science. In particular, he was apparently the first to represent an instantaneous rate of change by a straight line. The idea of an instantaneous velocity had been explicitly rejected by Aristotle. Indeed, even the idea of velocity as a simple ratio of distance to time was unacceptable to the Greeks, who confined the concept of ratio to like quantities. Bradwardine drew a distinction between velocity as a simple fraction of distance and time and as an instantaneous quality of motion.[3] In attempting to classify the concept of instantaneous velocity, Oresme stated that the greater this velocity, the greater would be the distance covered if motion were to continue uniformly at this rate. The Merton mathematician William Heytesbury made the same point.[4]

An important feature of the kinematic investigations by the Merton school and by Oresme was their discussion of the concept of acceleration, in particular uniform acceleration. In this respect, although not in the application to gravity, they anticipated Galileo by nearly three hundred years. Nevertheless, although they had the correct kinematic idea of time as the independent variable (which we now recognize as a fundamental feature of the scientific world-view), much confusion over this issue persisted until the time of Galileo. This confusion resulted from Aristotle's two definitions of the 'quicker', either as that which traverses a given space in less time, or as that which in a given time traverses a greater space. The former led to the erroneous conclusion that, in the naturally accelerated motion of falling bodies, velocity increased uniformly with respect to distance. This conclusion was supported even by Galileo before he arrived at the correct formulation.

Time and the mechanistic universe

As we have seen, the fourteenth century was a crucial period in the development of both the mechanical measurement and the mathematical theory of time. This was the century in which, owing to these and other technical innovations, we see the first signs emerging of the later predominance of West European civilization and thought.

One of the most influential ideas that can be traced back to that period is that the universe is a kind of machine. This was suggested by the elaborate astronomical clocks, for which a tremendous craze developed.[5] One of the most complex was the celebrated astrarium designed by Giovanni de' Dondi of Padua between 1348 and 1364.[6] This instrument was not merely a timepiece but a mechanical representation of the universe incorporating the celestial motions of the sun, moon and planets, including even the motion of the nodes of the moon's orbit, which takes over eighteen years to make a complete revolution around the ecliptic. In a treatise by Oresme, who was a contemporary of de' Dondi, on the question of whether the motions of the celestial bodies are commensurable or incommensurable, occurs an allegorical debate between Arithmetic, who favours commensurability, and Geometry, who favours the opposite. Arithmetic argues that incommensurability would detract from the harmony of the universe: 'For if anyone should make a mechanical clock, would he not move all the wheels as harmoniously as possible?'[7] This is the earliest instance known of the mechanical simulation of the universe by clockwork suggesting the reciprocal idea that the universe itself is a clock-like machine.

Nevertheless, it was not until the scientific revolution of the seventeenth century that this idea came to the fore. Kepler specifically asserted that the universe was like a clock, and Robert Boyle actually compared it with the famous clock at Strasbourg, 'where all things are so skilfully contrived that the engine being once set a-moving, all things proceed according to the artificer's first design.'[8] In the development of the mechanistic conception of nature in the course of the seventeenth century the mechanical clock played a central role. Moreover, the leading exponent of the mechanical philosophy in that century, Christiaan Huygens, was responsible for converting the mechanical clock into a far more precise instrument than it had been previously.

This development originated in Galileo's discovery of a natural periodic process that could be utilized for accurate time-keeping. As a result of much mathematical thinking on experiments with oscillating pendulums, Galileo concluded that each simple pendulum has its own type of variation depending on its length. The first pendulum clock was constructed by Huygens in 1656. It was based on his mathematical discovery that theoretically perfect isochronism could be obtained if the bob of the pendulum is constrained to move in a cycloidal arc.

Just as the introduction of mechanical clocks led to an improved system of time-reckoning, so the invention of a more precise time-measuring instrument had a great effect on the concept of time itself. For the construction of a clock which, if properly regulated, could tick away accurately for years greatly influenced belief in the homogeneity and continuity of time. This belief was implicit in the idea of time put forward by Galileo in the dynamical part of his *Two New Sciences*, published in 1638. Although he was not the first to represent time by a geometrical straight line, he became the most influential pioneer of this idea through his theory of motion, which was based upon it.

The idea of time as a mathematical concept that has many analogies with a line was first discussed explicitly in the *Lectiones Geometricae* (Geometrical Lectures) of Isaac Barrow, published in 1669. Barrow, who was Newton's predecessor in the Lucasian chair of mathematics in Cambridge, was greatly impressed by the kinematic method in geometry that had been so successfully used by Galileo's pupil Torricelli. To understand this method he concluded that it was necessary to study the concept of time. Barrow regarded time as essentially a mathematical concept that has many analogies with a line. He argued that 'time has length alone, is similar in all its parts, and can be looked upon as constituted from a simple addition of successive instants, or as from a continuous flow of one instant; either a straight or a circular line'.[9] The reference to 'a circular line' shows that Barrow was not completely emancipated from traditional ideas about cyclic time. Despite this, his statement goes further than any of Galileo's, for just as Euclid speaks only of straight-line segments and never of the complete straight line in our sense, so Galileo used only such segments to denote intervals of time.

Barrow's views on time greatly influenced Newton. In particular, Barrow's belief that, irrespective of whether things move or are still, time passes with a steady flow is echoed in the famous definition at the beginning of Newton's *Principia* (1687): 'Absolute, true and mathematical time, of itself and from its own nature, flows equably without relation to anything external.' Newton admitted that, in practice, there may be no such thing as a uniform motion by which time may be accurately measured, but he thought it necessary that, in principle, there should exist an ideal rate-measurer of time. Consequently, he regarded the moments of absolute time as forming a continuous sequence like the points on a geometrical line, and he believed that the rate at which these moments succeed each other is a variable that is independent of all particular events and processes.

Newton's views on time made a great impression on the philosopher John Locke. In his *Essay Concerning Human Understanding*, published in 1690, we find the clearest statement of the 'classical' scientific conception of time that was evolved in the seventeenth century:

Duration is but as it were the length of one straight line extended *in infinitum*, not capable of multiplicity, variation, or figure, but is one common measure of all existence whatever, wherein all things whilst they exist equally partake. For this present moment is common to all things that are now in being, and equally comprehends that part of their existence as much as if they were all but one single being; and we may truly say, they all exist in the same moment of time.[10]

Newton's belief that time is absolute was criticized by his contemporary Leibniz, who rejected the idea that moments of time exist in their own right. Instead, he thought of them as sets of events related by the concept of simultaneity, and he defined time as the order of succession of phenomena. Today this is generally accepted, and we regard events as simultaneous not because they occupy the same moment of time but because they happen together. We derive time from events and not vice versa. Nevertheless, Leibniz's definition concerns only the ordinal aspect of time and not its durational and metrical aspects.

Newton recognized the practical difficulty of obtaining a satisfactory measure of time. He pointed out that, although commonly considered equal, the natural days are in fact unequal. We now know that in the long run we cannot base our definition of time on the observed motions of the heavenly bodies. For the moon's revolutions are not strictly uniform but are subject to a small secular acceleration, minute irregularities have been discovered in the diurnal rotation of the earth, and so on. Instead, astronomers came to regard 'an invariable measure of time' as a measure that leads to no contradiction between the observed motions of the heavenly bodies and the rigorous theory of their motions.[11] In other words, the successes of celestial mechanics led ultimately to the conclusion that the clock-like behaviour of the universe is of less significance than that of the mathematical laws formulated by Newton to investigate it. This point of view was explicitly recognized in one of the great classics of Newtonian mechanics, the treatise on *Natural Philosophy* by Thomson and Tait, published in 1890. In discussing the law of inertia, they argued that it could be expressed in the following form: the times during which any particular body not compelled by force to alter the speeds of its motions passes through equal spaces are equal; and in this form they said that the law expresses our convention for measuring time.[12] Consequently, the conception of time that resulted from the successes of Newtonian mechanics was not, as Newton had himself imagined, of a physical entity existing in its own right. Instead, it was the universal mathematical time implied by his laws of motion.

Time and relativity

Until the beginning of the present century it was generally assumed that time was like a moving knife-edge covering all places in the universe simultaneously, and that the only arbitrary elements in its mathematical formulation were the choice of time unit and time zero. Apart from these, time was thought to be universal and unique. It was therefore a great surprise when, in 1905, Einstein discovered a previously unsuspected gap in the theory of time-measurement that led him to to reject the prevailing idea of time. In his analysis of the nature of the velocity of light, it occurred to him that time-measurement depends on the concept of simultaneity, the world-wide nature of which had been taken for granted. Einstein realized that, although this concept is perfectly clear when two events occur at the same place, it is not equally clear for events in different places. Instead, the concept of simultaneity for a distant event and one in close proximity to the observer is an inferred concept, depending on the relative position of the distant event and the mode of connection between it and the observer's perception of it. If the distance of an external event is known, and also the velocity of the signal that connects it and the resulting percept, the observer can calculate the epoch at which the event occurred and can correlate this with some previous instant in his own experience. This calculation will be a distinct operation for each observer, but until Einstein raised the question it had been tacitly assumed that, when we have found the rules according to which the time of perception is determined by the time of the event, all perceived events can be brought into a single objective time-sequence, the same for all observers. Einstein not only realized that it was a hypothesis to assume that, if they calculate correctly, all observers must assign the same time to a given event (provided that their clocks are properly calibrated), but he argued that, in general, this hypothesis must be abandoned. He believed that there are no instantaneous connections between external events and the observer. The classical theory of time, with its assumption of world-wide simultaneity for all observers, in effect presupposed that there are such connections.

Einstein was not only prepared to abandon the classical theory of time, but also the mechanistic conception of nature associated with it. He rejected all attempts to describe light and other forms of electromagnetic radiation as oscillations in a universal medium, the ether, for the mechanical properties that had to be assigned to the ether seemed to defy explanation. Instead, he adopted the principle of special relativity, which he regarded as fundamental. This principle asserts that all the laws of physics, and not just the laws of mechanics, are the same for all observers associated with inertial frames of reference. In particular, since the properties of light (and other forms of electromagnetic radiation) must be the same for all such observers, Einstein concluded that each must assign the same velocity to light in empty space.

Einstein found that, although the invariance of the velocity of light is compatible with the idea of world-wide simultaneity for all observers at relative rest, those in uniform relative motion will, in general, be led to assign different times to the same event, and that a moving clock will appear to run slow compared with an identical clock at rest with respect to the observer. The phenomenon of the apparent slowing down of a clock in motion relative to the observer is called 'time dilatation'. It is essentially a phenomenon of measurement applicable to all forms of matter. For two observers A and B in uniform relative motion it is a reciprocal effect: B's clock seems to A to run slow, and equally A's clock seems to run slow according to B. This reciprocity no longer holds, however, if forces are applied to change the uniform motion of one of the observers. In particular, if A and B are together at some instant, and are in uniform relative motion until at some later instant the motion of B is suddenly reversed so that he eventually comes back to A with the same speed, the time that elapses between the instant at which B left A and the instant when he returns to A will be shorter according to B's clock than according to A's. This result, which is significant only for speeds that are appreciable fractions of the velocity of light, means that if we accept special relativity we can no longer agree unconditionally with Isaac Barrow, who in his analysis of the classical concept of time argued that he did not 'believe there is anyone but allows that those things existed, equal times which rose and perished together'.[13]

In recent years much evidence in favour of the existence of time dilatation has come from the study of fast-moving particles. A good example concerns certain cosmic-ray phenomena. Elementary particles known as mu-mesons, found in cosmic ray showers, disintegrate spontaneously, their average 'proper lifetime', that is, time from production to disintegration according to an observer travelling with the mesons, being about two microseconds (two millionths of a second). These particles are mainly produced at heights of about ten kilometres above the earth's surface. Consequently, those observed in the laboratory on photographic plates must have travelled that distance. But in two microseconds a particle that travelled with the velocity of light would cover less than a kilometre, and according to the theory of relativity all material particles travel with a velocity less than that of light. However, the velocity of these particles has been found to be so close to that of light that the time-dilatation factor is about ten, which is the amount required to explain why it is that, to the observer in the laboratory, these particles appear to travel about ten times as far as they could in the absence of this effect.

Einstein's special theory of relativity is incompatible with Newton's concept of absolute time, but it can be regarded as a development of Leibniz's concept of time as the order of succession of phenomena. For although Leibniz himself envisaged a single time-system, the idea that time is derived from events, which is the essence of this theory, is compatible with the existence of a multiplicity of time-systems associated with different observers. Nevertheless, there is an important difference between Einstein's theory and Leibniz's as regards temporal order. For even if the temporal interval or duration of time between two events depends on the observer, we might still expect that, in special relativity, temporal order would be independent of the observer, since it is bound up with our ideas of causality. On the contrary, one of the most surprising consequences of special relativity is that, in certain circumstances, temporal order is dependent on the observer. If, according to an observer A, the distance r between two events E and F is less that ct, where t is the time-interval between them and c is the velocity of light, the temporal order of E and F will be the same for all observers. If however, r is greater than ct, uniformly moving observers can be found for whom the temporal order of E and F is the reverse of the order according to A.

The ostensibly paradoxical result that the temporal order of certain events can actually be reversed by an appropriate change of observer is closely associated with the restriction that, in special relativity, no causal influence can be transmitted with a speed exceeding that of light. For the events in question are precisely those that cannot be related by a signal unless it travels faster than light. On the other hand, the temporal order of events that can be causally related is the same for all observers in uniform relative motion.

Time and space-time

Besides the slowing down of the apparent rate of a moving clock, Einstein's special theory of relativity also predicts an apparent contraction of the length of a moving body, measured in the direction of its motion. This effect cannot be detected by an observer moving with the body, because all his instruments are affected in the same way, the contraction-factor depending on velocity relative to the observer. It had previously been suggested that a contraction of the same amount was needed to account for the failure of the Michelson-Morley experiment of 1887 to detect any motion of the earth with respect to the ether, although in the course of half a year the earth changes its velocity relative to the sun from 30 kilometres a second in one direction to the same speed in the opposite direction, and the apparatus used was thought to be sufficiently sensitive to detect a velocity of the earth relative to the ether of less than 10 kilometres a second. But the original suggestion of length-contraction, made independently by Fitzgerald and Lorentz to account for the null result of the Michelson-Morley experiment, implied that there were real structural changes in the constitution of matter due to motion relative to the ether. Einstein's theory involved no such complicated intrinsic changes but only an apparent contraction relative to the observer that accounted automatically for the Michelson-Morley result. Like the phenomenon of time-dilatation, length-contraction in relativity is a reciprocal effect: for two observers A and B in uniform relative motion, a body at rest with respect to A appears to B to be contracted in the same way as one at rest with respect to B appears to be contracted according to A.

Einstein's theory therefore implied that measurements of both space and time depend essentially on the observer. That this makes possible the forging of a much closer link between these concepts than had been previously envisaged was first realized by the mathematician Hermann Minkowski. His object, expounded in a famous lecture on 'Space and Time' delivered in 1908, was to replace the Newtonian concepts of absolute space and absolute time by a single new absolute four-dimensional concept, afterwards called space-time, and to show that its properties could be deduced from Einstein's theory. In Newtonian physics both the time-interval and the space-interval between two events are invariant, that is to say are independent of the observer. In special relativity neither is invariant, their values with respect to observers in uniform relative motion being subject to time-dilatation and length-contraction effects, respectively. Instead, Minkowski found that, if units of measurement are chosen so that the velocity of light is unity, then the difference between the square of the time-interval and the square of the space-interval is invariant. He therefore based his concept of space-time on the principle that this invariant represents the square of the space-time interval between two events. In Minkowski's four-dimensional space-time an event occurring at a point of space at a point of time is represented by a space-time point, and a particle, or other physical object, enduring for an indefinite time, is represented by a line, which Minkowski called its 'world-line'. Particles at rest or in uniform motion are represented by straight world-lines, and those in accelerated motion by curved world-lines. In this way, Minkowski reduced the study of motion to geometry.

Minkowski's concept of space-time has proved to be one of the most valuable contributions to physics made by a mathematician. It was used, in a generalized form, by Einstein when he later extended his theory of relativity to all types of motion. Because space-time is invariant, whereas its resolution into three-dimensional space and one-dimensional time varies from one observer to another, Minkowski declared: 'Henceforth space by itself, and time by itself, are doomed to fade away into mere shadows, and only a kind of union of the two will preserve an independent reality.' This celebrated claim must, however, be regarded as excessive. It tended to reduce the importance of time much more than of space. For in so far as it led those who accepted it to identify space-time with 'physical reality', it meant that the latter came to be regarded by them as 'a four-dimensional existence instead of, as hitherto, the evolution of a three-dimensional existence', to quote Einstein. In other words, many of those who looked upon the concept of space-time as a physical absolute, and not just as a useful mathematical tool, believed that the passage of time is a purely subjective feature of human consciousness, and they rejected the possibility of any objective, and therefore physically significant, time-scale of the universe. That this was too drastic a conclusion to draw we shall now see.

Time and the expanding universe

For a given observer, all events throughout the universe to which he assigns the same time define an instantaneous state of the universe. The order of succession of these states defines time as a whole for him. Consequently, whereas for Newton time was independent of the universe and for Leibniz time was an aspect of the universe, Einstein's theory leads us to regard time as an aspect of the relationship between the universe and the observer. Since Einstein's special theory of relativity, and also the general theory of relativity that he formulated some ten years later, imply that there is no universal time that is the same for all observers, it might be thought obvious that the idea of a cosmic time-scale can no longer have any objective significance, unless we abandon these theories. Such a conclusion would, however, be mistaken, because of developments that have taken place in cosmology.

According to current ideas, the universe is composed of galaxies, or systems of stars, broadly similar to our own Milky Way system. The spectra of galaxies exhibit red-shifts, which are thought by most astronomers to indicate recessional motion. These red-shifts appear to increase more or less systematically with distance. This has been taken as evidence that the universe as a whole is expanding, the rate of mutual recession of its parts increasing with distance. The discovery of this phenomenon by Hubble in 1929, now generally known as 'Hubble's law', made as great a change in man's picture of the universe as did the Copernican revolution, for previously the general assumption had been that the universe overall is static. Hubble's discovery stimulated much work in theoretical cosmology and drew attention to mathematical investigations made several years earlier by Friedmann and Lemaître, who independently had constructed expanding world-models. Other expanding models that were somewhat simpler were devised in 1932 by Einstein and de Sitter, and also by Milne. In all these models there are certain privileged or 'fundamental' observers, namely those fixed in the different galaxies. In general, the local times of these observers fit together to form one world-time, called 'cosmic time'. In other words, according to these observers, there are successive states of the universe as a whole that define a universal time. In terms of this, all events have a unique time-order, the anomalies and discrepancies of time-ordering and time-scale that arise in connection with relativity being due, not to the events themselves, but to the introduction of observers moving relative to the fundamental observers in their neighbourhood. From this point of view, the relativity of time is an essentially local phenomenon for observers in motion relative to the cosmic background defined by the bulk distribution of matter in the universe.

The existence of cosmic time was challenged by the mathematical philosopher Kurt Gödel in a paper published in 1949.[14] Although he agreed that the general distribution of matter and motion in the universe may offer a simpler aspect to some observers than to others, he did not believe that the aggregate of local times associated with such a class of privileged observers must automatically constitute a universal time. He constructed a homogeneous world-model in which the local times of the privileged observers who move with the galaxies cannot be fitted together into one world-wide time. Although his model was not entirely satisfactory, since it was theoretically possible for an observer moving sufficiently fast to make a complete circuit in time, a similar model free from this defect was discovered by Ozsvath and Schücking in 1962.[15]

An important difference between earlier homogeneous models of the expanding universe and these models concerns rotation. In constructing previous world-models it was assumed that any fundamental observer (associated with a galaxy) would see himself to be at a centre of isotropy or spherical symmetry, so that there would appear to be no preferential directions in the universe. Consequently, in these models there are no systematic cosmic motions transverse to the line of sight and hence no cosmic rotation. The absence of cosmic rotation means that at each point the directions of cosmical recession are like the spokes of a stationary wheel, except that they occupy three-dimensional space and do not lie in a plane. Moreover, these directions coincide with the directions of inertial motion (free motion not subject to local forces), which constitute what has been called 'the compass of inertia' at each point. In the world-models constructed by Gödel and by Ozsvath and Schücking the system of galaxies is observed from each point to rotate relative to the local compass of inertia.

It has been established mathematically that any world-model that is isotropic about each fundamental observer is homogeneous and is characterized by a cosmic time.[16] It follows that, to help in deciding whether there is a cosmic time for the actual universe, we must look for empirical evidence for isotropy beyond that furnished by the somewhat patchy distribution of the galaxies. Impressive support for the assumption of world-isotropy has, in fact, come from the discovery in 1965 of cosmic microwave radiation.[17] Unlike starlight, this radiation is more or less isotropic, the variation in intensity with direction being about 0.1 per cent. This degree of isotropy excludes the possibility of the origin of this radiation being in the solar system, our galaxy or the local cluster of galaxies. This radiation is therefore believed to be a constituent of the universe as a whole, and the black-body character of its spectrum indicates that it is a relic of the intense primeval radiation that may have accompanied an explosive origin of the universe some 10,000 to 20,000 million years ago. Since any large-scale departures from homogeneity and isotropy in the universe would affect the radiation and make it seem anisotropic to us, we have powerful evidence that the universe as a whole is predominantly homogeneous and isotropic, and hence that there is a cosmic time.

Time and world-horizons

John Locke in his *Essay Concerning Human Understanding* concluded a chapter on space and time by declaring that 'expansion and duration do mutually embrace and comprehend each other; every part of space being in every part of duration, and every part of duration in every part of expansion'.[18] This statement was not disputed, for it seemed a truism, until in 1917 de Sitter constructed a static world-model in which time was subject to a previously unsuspected limitation. In the experience of an observer located at a given point in the model, there was a finite horizon at which time appeared to him to stand still, as at the Mad Hatter's tea party where it was always six o'clock. An observer located on this horizon would, however, experience the normal time-flux. The effect occurred because the time required by light or any other electromagnetic signal to travel from an observer's horizon to himself was infinite.

Nowadays it is customary to choose co-ordinates of space and time so that de Sitter's world is regarded as an expanding universe with a cosmic time, its rate of expansion being an exponential function of this time. In terms of cosmic time, if *A* and *B* are any two fundamental observers in the model, there will be an epoch in the history of *B* that will appear to *A* to be a time-horizon in the sense that no signal emitted by *B* at that epoch or later can ever reach *A*. Similarly, according to *B*, there will be a corresponding time-horizon associated with *A*. A time-horizon of this type is called an 'event-horizon'. It will exist for any fundamental observer *A* in any expanding world-model for which the rate of expansion increases so rapidly that eventually no signal emitted by another given fundamental observer *B* can ever arrive at *A*. From *A*'s point of view, light from *B* is then, in Eddington's graphic phrase, 'like a runner on an expanding track with the winning post receding faster than he can run'.[19]

A different type of time-horizon exists in any world-model that has a decreasing rate of expansion that was at first so fast that no signal emitted by a particular fundamental observer *B* could reach the fundamental observer *A*. In other words, only after a certain epoch can a particular observer *B* emit a signal that will eventually arrive at a particular observer *A*. This type of time-horizon is called a 'particle-horizon', because it seems to each fundamental observer *A* that matter is continually coming into existence at the confines of the visible universe.

Most of the expanding world-models that have been studied possess one or other type of horizon, or even both. A notable exception is the model due to Milne, in which the rate of expansion is uniform. This model possesses neither type of horizon, and so maintains a more complete unity in time than many other models. In general, however, the concept of a fundamental cosmic time, associated with the idea of successive states of the universe, is limited by the existence of time-horizons. Moreover, even if the observer is no longer assumed to be located in a particular galaxy but moves through the universe, with a local speed less than that of light, his time-horizon may change but can never be abolished. If the model possesses an event-horizon for any fundamental observer, then no observer moving through the model, with any local speed less than that of light, can observe every event in the universe.

Time-scales and universal constants

The existence of a world-wide scale of time does not necessarily imply that for a given observer there is a unique scale of time in nature. For if we adopt a relative measure in terms of a particular sequence of natural phenomena—for example, the successive swings of a pendulum—we may obtain a scale of time that is adequate for the temporal ordering of all phenomena but is not necessarily adequate for the metrical comparison of all different intervals of time. Given three successive events E, F and G, at the same place, the temporal intervals between E and F and between F and G, respectively, might appear to be of equal duration according to a pendulum clock, and of unequal duration according to a different type of clock—for example, one associated with some atomic process. If one clock were represented mathematically by the variable t and the other by the variable τ, the order in which events occur, locally at least, would be the same according to either clock so long as τ is a monotonically increasing function of t. If however, events that are periodic on the t-scale (for example, if the interval FG is equal to the interval EF), are to be also periodic on the τ-scale, then in general τ must be restricted to being a linear function of t. In that case, t is a linear function of τ and all events that are periodic on the τ-scale are also periodic on the t-scale. Two scales of time used by the same observer can therefore be regarded as effectively identical, except for possible differences in the choice of time unit and time zero, if and only if they are linearly related.

The French mathematician and philosopher of science Henri Poincaré argued that different ways of defining time would lead to different 'languages' for describing the same experimental facts. He concluded that time should be so defined that the fundamental laws of physics, in particular the equations of mechanics, are as 'simple' as possible. He claimed that 'there is not one way of measuring time more true than another; that which is generally adopted is only more *convenient*. Of two watches, we have no right to say that one goes true, and the other wrong; we can only say that it is advantageous to conform to the indication of the first'.[20] Poincaré appears, however, to have overlooked the possibility that the customary 'simple' formulations of distinct physical laws may entail different scales of 'uniform time', i.e. scales that are not linearly related. For we have no prior guarantee that the time-scale defined, for example, by the law of radioactive decay is the same as that implied by the laws of celestial mechanics.

The way in which this problem has usually been discussed is in terms of the possible variation in time of the values of one or more of the fundamental constants of physics. For, if there were no natural time-scale in terms of which all these 'constants' are secular invariants, it would mean that more than one scale of time would be required if we wished to work with universal constants that do not change with time. Although there has not been until recently any convincing evidence for discarding the hypothesis of a unique cosmic time in terms of which all the fundamental constants and laws of physics are independent of epoch, T. C. Van Flandern, of the United States Naval Observatory, claims that a comparison of the timing of occultations of stars by the moon using Atomic Time (based on the caesium atom) and the timing based on the apparent annual motion of the sun about the earth gives a discrepancy that could be attributed to a decrease in the universal constant of gravitation of the order of one part in 10^{10} per annum.[21] If this result is confirmed by others it will have far-reaching consequences, since a varying constant of gravitation is not compatible with general relativity. On the other hand, an experiment designed by W. A. Baum and R. J. Nielsen to compare the energy of old photons from remote galaxies with that of young photons from nearby galaxies indicates that Planck's constant does not change significantly with time, since old photons of a selected wavelength are found to have the same energy as young photons of the same wavelength.[22]

Conclusion

The discovery of the expansion of the universe has reinforced the tendency in recent centuries for time to become the dominant feature of the scientific world-view. Moreover, although the principle of relativity has undermined the concept of a single universal time-scale, the measurement of time has become increasingly significant in physics. This is due not only to the importance of effects such as time dilatation but also to the use of the radar principle, according to which distance is measured by the time taken by light (or other electromagnetic signals) to traverse it. In recent years the accuracy of time measurement has become remarkably precise, the caesium atomic clock producing a frequency standard that is accurate to two parts in ten million million, corresponding to a clock error of only one second in 150,000 years.

We have moved farther away than ever from the world-view of ancient science in which time was a concept of secondary importance compared with place and form. Today the idea of time as linear advancement without cyclic repetition has developed into a sophisticated mathematical concept that has both influenced, and been influenced by, the growth in our knowledge of the physical universe.

To conclude, let us briefly consider whether recent developments in the mathematical concept of time have exerted any significant influence outside science. Since the advent of modern industrial civilization, men's lives have come to be dominated more and more by the clock, but this has tended to make time appear as something absolute. Consequently, the relativistic conception of time, which has proved to be so important in physics, has had little effect on most people's idea of time. On the other hand, the division of the earth's surface into separate time-zones, and the adoption of summer time (daylight saving) in many countries, have made many aware of the conventional factors in time-reckoning. Nevertheless, it is not surprising that the more sophisticated aspects of the modern mathematical concept of time have not yet made any impact on mankind in general.

NOTES

1. L. Thorndike, *A History of Magic and Experimental Science*, Vol. 3, p. 290, New York, Columbia University Press, 1934.
2. L. Mumford, *Technics and Civilization*, p. 14, London, Routledge & Kegan Paul, 1934.
3. L. Lamar Crosby, *Thomas of Bradwardine, his Tractatus de proportionibus*, p. 44, Madison, University of Wisconsin Press, 1955.
4. Curtis Wilson, *William Heytesbury: Medieval Logic and the Rise of Mathematical Physics*, p. 21, Madison, University of Wisconsin Press, 1956.
5. Lynn White, *Medieval Technology and Social Change*, p. 124–5, Oxford, Clarendon Press, 1962.
6. S. A. Bedini and F. R. Maddison, 'Mechanical Universe: the Astrarium of Giovanni de Dondi', *Transactions of the American Philosophical Society* (Philadelphia), Vol. 56, Part 5, 1966.
7. E. Grant, *Nicole Oresme and the Kinematics of Circular Motion*, p. 295, Madison, University of Wisconsin Press, 1971.
8. T. Birch (ed.), *The Works of the Honourable Robert Boyle*, Vol. 3, p. 405, London, 1772.
9. I. Barrow, *Lectiones geometricae* (trans. E. Stone), Lect. 1, p. 35, London, 1735.
10. J. Locke, *Essay Concerning Human Understanding*, 1690, Book II, Chapter 15, Paragraph 11.
11. G. M. Clemence, *The American Scientist*, Vol. 40, 1952, p. 267.
12. W. Thomson and P. G. Tait, *Natural Philosophy*, Part 1, p. 241, Cambridge, Cambridge University Press, 1890.
13. I. Barrow, *Lectiones geometricae* (trans. E. Stone), Lect. 1, p. 5, London, 1735.
14. K. Gödel, *Reviews of Modern Physics*, Vol. 21, 1949, p. 447.
15. L. Ozsvath and E. Schucking, *Nature*, Vol. 193, 1962, p. 1168.
16. A. G. Walker, 'Completely Symmetric Spaces', *Journal of the London Mathematical Society*, Vol. 19, 1944, p. 219.
17. A. A. Penzias and R. W. Wilson *Astrophysical Journal*, Vol. 142, 1965, p. 419.
18. J. Locke, *Essay Concerning Human Understanding*, 1690, Book II, Chapter 15, Paragraph 12.
19. A. S. Eddington, *The Expanding Universe*, p. 73, Cambridge, Cambridge University Press, 1933.
20. H. Poincaré, 'The Value of Science', in *The Foundations of Science* (trans. G. B. Halsted), New York, The Science Press, 1929, p. 228.
21. T. C. van Flandern, 'A Determination of the Rate of Change of G', *Monthly Notices of the Royal Astronomical Society*, Vol. 170, 1975, p. 333–42.
22. W. A. Baum and R. F. Nielsen, *The Constancy of Planck's Constant*, Preprint, 1975.

TIME AND ASTROPHYSICS: THE TIME ARROW AND THE EXPANDING UNIVERSE

S. L. Piotrowski

The cosmos

Looking at the sky on a clear night and seeing myriads of stars shining on its dark background, we seldom realize that the fact that the night sky is dark reflects one of the most fundamental properties of time, namely the direction of its flow—i.e. the direction of 'the time arrow' (Bondi, 1962). It is relatively difficult to find physical processes that would demonstrate the lack of reversibility of time. No laws of physics hitherto discovered contain any indication of the direction of the flow of time: they do not change when the direction of the flow is reversed. Only in elementary-particle physics, the so-called $K_s^\circ - K_L^\circ$ transition gives an indication of the breaking of the time-reversal invariance.

There are two processes indicating directions of the flow of time: in thermodynamics the increase of entropy in an isolated system and, in Maxwell's theory of electromagnetic radiation, the necessity of considering retarded potentials (also, from a formal point of view, the advanced potentials).

In the case of the retarded potentials, their choice is imposed by the principle of causality: the retarded potential, in a given point of space and in a given moment of time, is determined by the state of the source of the electromagnetic field at the moment that precedes the considered moment by an amount of time required for the propagation of the field perturbation from the source to the given point.

Direction of the time flow connected with the principle of entropy increase results from statistical considerations. Partly, this is the meaning (but only partly, since it is deeply connected with the cosmological problem) behind the statement made at the beginning of this article about the connection between the darkness of the sky and the direction in which time is moving.

Suppose we could put a star between impenetrable walls, isolating it from the rest of the universe; then, after some time inside this enclosure, we could tell nothing about the direction of the time flow: the walls would reflect the stellar radiation, a great density of radiative energy would accumulate and a state of thermodynamic equilibrium would eventually prevail. The whole system would forget in which direction time goes. If we were now to make a small window in this box in which the star was enclosed, then the outflow of radiation would be very violent, while the inflow of outside radiation would be only very small—since the sky is dark. If we now closed the window again, the star would re-radiate more energy than the enclosure was sending back, and at least for some time the direction of the time arrow would be evident.

As I mentioned earlier, the darkness of the sky is intimately connected with the cosmological problem. In the eighteenth century Edmond Halley asked himself why the night sky is dark. Let us imagine a static, infinite universe that is filled uniformly with stars and galaxies. In every direction, the line of sight will eventually hit the surface of some star. Since the surface brightness of the stellar disc is independent of its distance, the whole sky should seem to be as bright as the surface of the average star (approximately as bright as the sun's surface). No such phenomenon is observed. This paradox, named after the German astronomer Heinrich Olbers, living at the turn of the eighteenth and nineteenth centuries slowed down the development of cosmology for a good many years. It is impossible to imagine a finite universe and hence the division of space into two parts: one where there is something and another where there is nothing.

It was not until 1929 that the paradox of Olbers was resolved. In that year the American astronomer E. P. Hubble showed that the universe is expanding, all galaxies are receding, and the velocity of recession is proportional to the distance of the galaxy, i.e. the larger the distance the greater the velocity. The recession of galaxies strongly affects the radiation reaching the earth, and does so in two ways: each successive photon coming to earth has a longer distance to travel, and consequently the rate of photon arrival is lower than in the case of a stationary galaxy. Moreover, because of the Doppler effect, the colour of photons is shifted to the red and they are less energetic. The dimming of light caused by both these effects (their effectiveness is greater for more distant galaxies) leads to the phenomenon known to all of us—that the night sky is not as bright as the sun's disc but dark. As we see, the basic fact connecting the time-symmetrical laws of physics with our intuition and experience about unidirectional time flow is supplied by the astronomical evidence. Though the evidence is quite straightforward—the night sky is dark—it was understood only in recent decades. Let us add that our understanding is far from complete. We cannot, for instance, explain at all why the universe is expanding and what initiated its expansion—and such questions can be multiplied.

The earth as the astronomical clock

Probably the best definition of time is given by Aristotle: it is the quantity referring to motion from the point of view of earlier or later: *Touto gar estin o chronos arithmos kinēseos kata to proteron kai ysteron.* And in fact whenever we measure time we have to do it, in principle, by considering some motion: be it a clock-hand or the earth revolving around its axis, or circling the sun, or the electron orbiting around the nucleus. For long centuries the astronomers had a monopoly of measuring time. Its flow was defined by the rotation of the earth. The calendar was based (and still is today) on expressing in days the time required by the earth to make one revolution around the sun. From antiquity to the beginning of the modern era that measurement was not an easy one for astronomers. And it was only after long discussions during the Fifth Lateran Council and the much later decision of Pope Gregory XIII that the relatively good repeatability of the beginning of the year's seasons on the calendar's dates was reached. The operation of introducing the Gregorian calendar entailed omitting ten days between 4 and 15 October. It is worth mentioning that during the Lateran Council discussions in 1514 Copernicus was consulted as an expert on the calendar reform. It seems that the Gregorian reform of 1582 (O'Connell, 1975) was based on tables (*Tabulae Prutenicae*) that included Copernicus's results. It is comforting to note that although the Church strongly opposed the Copernican cosmology, it nevertheless used Copernicus's calculations in the calendar reform it sponsored. Perhaps we should always rely on the opinion of a good expert.

It appears today that, because of the lack of uniformity of motion, the rotating earth is not a very reliable clock. This has been shown with ultra-precise atomic clocks based on the frequency of a certain line of caesium. The discrepancies between the length of a day measured by the earth's revolution and that measured by an atomic clock amount to a few thousandths of a second. They result probably from changes of the moment of inertia of the earth, which in turn arise from the moments of the large masses of the earth's atmosphere. According to Burkard (1972), there even exists a correlation between solar flares and changes of the length of a day. These changes seem to be connected with those induced by the solar wind in the earth's magnetosphere and ionosphere.

From obvious considerations pertaining to the needs of everyday life, navigation, geophysics, etc., we must nevertheless use terrestrial time (as a rule it is the time of the mean Greenwich meridian). But this time is co-ordinated in such a way as to make the difference between it and atomic time an integral number of seconds. We have therefore to add or subtract one second now and then. Since we are dealing here only with seconds, we do not need the Pope to intervene every time. Let us add that astronomers, apparently distrustful of the atomic clock invented by the physicists, also use a so-called ephemeris time, which is defined as the time in which the motion of celestial bodies (in the first place the sun) is governed by the laws of Newtonian mechanics. It is only on the basis of observations of these celestial bodies (hence with considerable delay) that the difference between ephemeris time and Greenwich time is determined. On 1 January 1975, the difference between atomic time and Greenwich time was 14 seconds, while that between ephemeris time and Greenwich time was as much as 45 seconds.

Perhaps the progress of ephemeris time, determined by the planetary motions, is distorted in a peculiar way. Some recent measurements (Flandern, 1975) seem to indicate that Newton's force of gravitation changes in time: it gets weaker because the proportionality factor in Newton's law decreases at the rate of one part in 10^{10} per year.

The age of the earth and geology

We have dwelt upon methods of measuring time because today it is the only domain in which astronomy intervenes directly in everyday practice. On the other hand, it is not seconds or days that are of importance for most of the natural sciences and our culture, but very long intervals of time in which contemporary astrophysics sees or foresees evolutionary changes: of our sun, of stars, of the whole system of stars of the Milky Way to which our sun belongs, and of the whole universe. Let us not forget that almost until the seventeenth century the longest time scale in Europe was the scale of the Bible—4,000 years. The world of the stars was invariable. The famous astronomer Tycho Brahe, when writing in 1572 about the nova observed by him, wondered: 'All philosophical conceptions indicate . . . that the sky and the celestial bodies do not grow nor get smaller, do not undergo any changes neither in respect to number, to appearance, brightness.' In fact, it was only geology and the discoveries of palaeobotany and palaeozoology that discerned the really long periods of time that were essential from the evolutionary point of view. This was not before the nineteenth century. Darwinian evolutionary theory (1859) permitted the arrangement into time sequences of sedimentary rocks and fossils. Nevertheless, it was not possible to determine the time scale in years. This became possible only with the discovery of radioactivity. The age of different minerals could then be determined from the abundance of disintegration products (in the first place, uranium). The first determinations obtained by this method, before the First World War, pointed to the age of the earth as being of the order of 1,000 million years (10^9 years). But during the whole second half of the nineteenth century and the beginning of the twentieth century, much shorter estimations of the earth's age were common: estimates of the order of a few hundred million years. These calculations were based for instance on estimates of the time required for the earth to cool to the observed temperature, or of the time needed for the formation of sediments of the observed thickness. It is worth mentioning that by 1901 Maurycy Rudzki, professor at the Jagellonian University in Cracow, had already given an estimate of the earth's age quite close to that held currently. He assumed that mountain ranges arise because the earth's radius contracts during the process of cooling. In this way he obtained an age for the earth of 3,000 million years (3×10^9 years).

Geology is in essence a history of the earth. As already mentioned, geological chronology is based on evolutionary changes in flora and fauna in the course of centuries. Although the radioactive-decay method is essential in determining the absolute age of the rocks, it is the evolution of organic life that identifies the large periods in the earth's history. The oldest living cells had already appeared on the earth 3,000 million years ago, but for some 2,000 million years only very primitive organisms existed. The development of organic life was probably limited by the lack of oxygen in the earth's atmosphere, and by the deadly influence of ultraviolet rays, which were not yet blocked by any screen of ozone. The sudden development of organic life on the earth took place some 600 million years ago, and was probably connected with the evolution of the earth's atmosphere, in which the minimum amount of oxygen necessary for breathing appeared. The next threshold in the development of organic life on earth was passed when, some 400 million years ago, the amount of oxygen in the atmosphere reached the level necessary for the formation of an ozone screen barring the penetration of deadly ultraviolet light (the oxygen content still amounted then to only 10 per cent of the present value).

It is this very possibility of the depletion of the ozone screen by the combustion products of supersonic jets that is often invoked as an argument against the development of such aircraft.

A very convenient clock for timing the not-too-distant geological past is offered by certain astrophysically conditioned processes. The earth's atmosphere is constantly being bombarded by cosmic rays. This is the radiation, whose primary composition (primary, i.e. prior to entering the atmosphere) consists mainly of protons and alpha particles. The cosmic rays passing in the vicinity of the earth are mainly of galactic origin, with only a small number emitted by the sun. Primary cosmic rays interact with particles in the atmosphere, make them disintegrate, and create cascades of secondary particles, among which are some neutrons. These neutrons colliding with nitrogen atoms transform them into a radioactive isotope of carbon, ^{14}C, which in turn binds two atoms of atmospheric oxygen and forms carbon dioxide. The carbon dioxide diffuses into the lower layers of the atmosphere, is assimilated by plants, and through them by animals: every living organism contains radioactive carbon. The radio-carbon decays with a half-life of 5,600 years. In living organisms the decaying radio-carbon is constantly replenished, and the ratio of the content of radio-carbon to the 'ordinary' ^{12}C is kept constant. If the organism dies, the replenishment of ^{14}C stops and the radioactivity of organic remains diminishes at a rate strictly determined by the law of radioactive decay. In this way we can accurately establish the time that has elapsed since the death of an organism. Unfortunately, because of the short life-span of radio-carbon, our ^{14}C clock can reach back only into the not-too-distant past: a few tens of thousands of years.

We can control the accuracy of this clock with the help of some historically established dates, e.g. the dates of the deaths of Pharaohs in ancient Egypt. The age of the wood from which the coffins and the burial boats were made, as determined by the ^{14}C clock, is compared with their age as assessed by historians. It turned out that the radio-carbon clock needs some corrections. Apparently at times radio-carbon was produced and therefore assimilated more intensively. This irregularity of our clock was discovered, and the necessary recalibration introduced, by direct counting of annual tree rings of growth in very old specimens of *pinus aristata*.

The ^{14}C clock permits us to find the dates of birth and the stages of development of various prehistoric cultures in different parts of our globe. In particular, we can decide which culture was prior to the other and therefore which could have been influenced by the older one. Such absolute time ascertainments were often impossible before the radio-carbon clock was discovered. It is fascinating to realize that the cosmic rays from distant stars and galaxies generated the time-measuring mechanism for facts of the history of culture in ages when no written history existed.

Absolute dating in geology based on radioactive decay now often relies on the relative content of radioactive potassium ^{40}K and argon in minerals. The natural isotope ^{40}K decays into argon. This method allows us to go back some 50 million years. Probably the most accurate method of this type consists in counting traces of disintegration of individual uranium atoms in minerals.

The thermal and nuclear time-scales of contemporary astrophysics

Moderately long geological periods (compared with the longest ones, of the order of thousands of millions of years), that is to say, periods of a few hundred million years, were much too long for nineteenth-century astrophysics. The sun is the source of life on the earth, and nineteenth-century astrophysics did not know how to make it shine long enough. The only acceptable process was the generation of the sun's thermal energy contraction, a mechanism proposed by Helmholtz in the 1850s. The upper limit of the sun's age, even if it contracted from infinity, and shone always as it does today, would come to only a few tens of millions of years. Fossil flora and fauna clearly testify that the sun has had to radiate in much the same way as today for at least 1,000 million (10^9) years. The difficulty of reconciling the two time-scales (1,000 million years as against a few tens of millions of years) was resolved definitively just before the Second World War. In 1938, the American physicist Hans Bethe succeeded in explaining the riddle of the sun's energy source in terms of nuclear fusion in which four nuclei of hydrogen are condensed into one nucleus of helium. About 0.7 per cent of the total hydrogen mass is converted into energy. The theory of the internal structure of stars indicates that this amount of energy is sufficient to cover the energy expenditure for a star such as the sun for many billions of years.

Perhaps we will soon be once again in the situation of astronomers at the end of the nineteenth century and the beginning of the twentieth, when no answer was available to an important question. The question no longer is 'Why are the sun and other stars radiating so strongly?' but 'Why do they emit energy for such long periods of time?' We are now wondering why some objects are radiating so strongly, and may be neglecting the fact that this paltry 0.7 per cent of the mass may turn out to be too small with respect to the observed expenditure of energy. I have in mind the quasars, some radio galaxies and violent events observed in the nuclei of certain galaxies and perhaps in certain stars (Chandrasekhar, 1972).

The rotation of our galaxy
and the ice ages in geology

Without escaping to the remote galaxies let us concentrate for a while on our own galaxy, the Milky Way. New data from the branch of theoretical astrophysics that deals with the theory of the structure of the galaxy, and some results obtained from the lunar missions, seem to indicate that we can today link processes going on in the galaxy with climatic processes on the earth—namely, with the ice ages. Just before the Second World War, in 1939, Hoyle and Lyttleton suggested that the passage of the solar system through dense clouds of interstellar matter could result in an ice age on the earth. Although it seems paradoxical, it is through accretion of the interstellar dust that the sun gets a bit warmer and radiates more energy to increase its albedo and so initiates an ice age. The spiral arms observable in other galaxies and which also exist in our Milky Way are regions with many bright stars and interstellar gas and dust. A theory developed in the 1960s by Lin (1964) explains that the observed spiral structure is a wave pattern of stationary character through which the stars and interstellar matter move. These objects enter the spiral arm through its inner concave edge. A shock wave developing along this edge leads to a condensation of clouds of interstellar matter and something like a dust lane. It is there that the stars are born first. In such a compressed cloud some 5,000 million years ago, the sun was born. One can calculate that the sun revolving around the centre of the galaxy crosses a spiral arm once every 100 million years, and for a million years is crossing the compression lane. These periods (an ice age once every 100 million years and glaciation lasting 1 million years) are confirmed by geological evidence. This theory, presented by McCrea (1975), was unexpectedly supported by the studies of Lindsay and Srnka (1975) on the erosion of lunar soil caused by streams of micrometeorites striking the moon's surface. From the material brought back by the Apollo XV mission, it appears that the intensity of the flux of micrometeorites on the moon changes to the same pattern as the ice ages on earth. The evidence is still not conclusive, but we have here an excellent example of long-lasting multi-million-year astronomical cycles, just recently discovered, influencing the processes developing on the earth's surface.

The astronomical and the physical age of the galaxy

In the galaxy and on its outskirts, there are some compact agglomerations of stars called globular clusters. It can be reasonably assumed that the stars belonging to such clusters were born at the same time from matter of the same chemical composition. Accepting such assumptions, the theory of the internal structure of stars leads to the conclusion that the way in which the brightness of the stars belonging to a given cluster changes with their temperature is strongly dependent on the age of the cluster. Thus, having determined from observation the shape of the above-mentioned relation (brightness versus temperature), the age of the cluster can be assessed. The time intervals obtained in this way are of the order of 10,000 million (10^{10}) years. Globular clusters appear to be the oldest components of the galaxy. The chemical composition of stars belonging to globular clusters gives similar evidence. This composition is characterized by a deficiency of metals and is probably identical with the composition of matter as it was at the very birth of the galaxy. In the later stages of the galaxy's life, the matter from which the second-generation stars were formed was enriched in heavy elements—metals, ejected from the first-generation stars.

This 'astronomical' age of the galaxy, determined by the galaxy's oldest components (globular clusters), can be confronted with the evidence supplied by the history of nucleosynthesis within the galaxy. The contemporary theory of nuclear synthesis in the galaxy seems to indicate that the heaviest elements in the galaxy were born in the first 100 million years of its existence. Hence, if we take a pair of two heavy radioactive nuclei with different decay rates (and thus different half-lives), e.g. ^{235}U and ^{238}U isotopes, or elements ^{232}Th and ^{238}U, and compare their present, observed, relative frequency with the expected relative frequency (of the order of one) at the moment of the creation of the galaxy, we can compute, using their known life-times and the law of exponential decay, how much time has elapsed from their creation in the first stage of the life of the galaxy. This is the 'physical' age of the galaxy. This age comes out somewhat below 10,000 million years (Trimble, 1975). Taking into account both the uncertainty of the observational data and an uncertainty inherent in the models, we can consider both ages, astronomical and physical, as roughly consistent.

The longest astrophysical time-scale

The longest astrophysical time-scale is that connected with the problems of the evolution of the whole universe. Contemporary astrophysics sees this universe as being composed of galaxies (similar to our own galaxy, the Milky Way), clusters of galaxies, a very rarefied gas (mostly hydrogen) dispersed among them, and radiation. Since Hubble's time we know that the universe is expanding; galaxies recede, with a velocity of recession proportional to their distance. In the relation expressing the proportionality of the velocity of recession V to the distance r

$$V = H.r$$

the coefficient of proportionality, H, is called Hubble's constant. It can be seen that the reciprocal of H, $1/H$, can be regarded as a time interval needed by a galaxy to cover the distance from the initial value, formally equal to zero, to its actual value, r. We can look at $1/H$, rather naively, as a time in which the whole universe expanded from a certain small initial volume to its present dimensions.

The present data indicate that the value of Hubble's constant is well below $100 \text{ km s}^{-1} \text{ Mpc}^{-1}$. With Hubble's constant equal to $65 \text{ km s}^{-1} \text{ Mpc}^{-1}$, the time of the expansion of the universe comes to about 15,000 million years. This is really a cosmic time scale: the interval between the present and some initial moment at which the expansion started.

The hot 'big bang' and the background radiation

Besides data on the velocities of recession of more and more distant galaxies, the last decade has provided another piece of observational information of extreme importance. This information permits us to extrapolate back, with a considerable degree of confidence, in the evolution of the universe. This is the discovery of the so-called background radiation of a very great degree of isotropy (Penzias and Wilson, 1965). The sea of radiation which permeates the whole universe embraces radiation of different frequencies: from gamma rays to radio waves. We are concerned here with background radiation in the microwaves range, from a few dozens of centimetres to a fraction of a millimetre. The energy spectrum of this microwave radiation corresponds, with a very good accuracy, to that of the radiation of a black body with a temperature a little below 3° K. In the canonical theory of the hot 'big bang', which I am presenting here, the interpretation of this fact is quite clear. It is fascinating to note how precisely, relying on the known principles of physics, we can follow details of the most distant past of this world of ours, starting from observational data about the recession of galaxies and the temperature of microwave background radiation.

Let us move back in time by roughly 15,000 million years, to the first seconds of expansion of the universe (Peebles, 1971; Webster, 1974). Initially, the temperature and pressure were so high that speculations as to what particles interacted with photons are rather arbitrary. At the moment when the temperature dropped to about ten million million (10^{12}) degrees, we can assume that the only particles that existed in equilibrium with photons were electron-positron and neutrino-antineutrino pairs.

Sometime at the moment when the temperature dropped to about 1,000 million degrees—that was a hundred seconds after the 'big bang'—the matter, still of the same temperature as the radiation, was composed mainly of nuclei of hydrogen and helium, and of course of electrons, neutrinos and antineutrinos. Around this time, most of the helium existing now in the universe was created. This allows some possibility for the observational testing of the theory of the 'big bang'. As the universe expanded, the temperature continued to decrease, and around 10^{13} seconds after the 'big bang', at a temperature of about 5,000° K, electrons recombined with ionized gas emitting radiation—the last relic of the initial fire-ball—which, shifted by the Doppler effect to the microwave range, is observed today as the mentioned background radiation of 3° K. This recombination happened barely some hundred million years after the 'big bang'. Only 2,000 million years after the 'big bang', the galaxies started to form.

Is the universe open?

Will the universe always expand, or will it oscillate cyclically—in other words, using the language of the general theory of relativity, will there thus be a closed or an open universe? This depends on the value of a certain parameter q, called the parameter of deceleration, which measures the speed of changes of the expansion, normalized in a certain way, and is proportional to the density of matter in the universe (and inversely proportional to the square of Hubble's constant). The values of q larger than $1/2$ correspond to a closed universe; the values of q smaller than $1/2$ to an open, uniformly expanding universe. It is still not adequately explained (assuming we know Hubble's constant with sufficient accuracy) whether the density of matter in the universe is enough to give it a closed character. Some recent works (Got III, Gun, Schram, Tinsley, 1974; Sandage, 1975) seem to indicate that q is smaller than $1/2$. May I add that some analyses of Hubble's diagram (illustrating the dependence of the velocity of recession of galaxies on their distances), performed in Warsaw (Kruszewski, 1976, unpublished) point to the inadequacy of present data for inferring the open character of the universe.

Astrophysics and the space-time continuum of the general theory of relativity

Speaking about the model of the expanding universe above, I used the phrase 'the language of the general theory of relativity'. The word 'language' is perhaps not the right one. It is not a matter of language, but of a way of looking at time and space, fundamentally modified by the general theory of relativity. The theory of relativity related time and space and made us look at the world as a four-dimensional continuum, and made us see the laws of motion of matter as resulting from space-time geometry. This theory was in the deepest sense of the word a breakthrough, not only for our knowledge of the cosmos, but for the whole of physics. I would be prepared to say that the theory of relativity impressed upon scientists an evolutionary look at the universe. This theory found in astronomy its best testing ground. All recent experiments on the validity of the general theory of relativity concern the behaviour of electromagnetic waves in the vicinity of massive objects, and require high-precision measurements of small angles or very short time intervals.

When radar signals reflected from Mercury or Venus or signals emitted by a cosmic probe pass near the sun, they should suffer a retardation of the order of a few ten-thousandths of a second. Such retardation was actually observed in several experiments, and its magnitude agreed with the prediction of the general theory of relativity.

The late stages of the evolution of stars and black holes

An ever-improving understanding of nuclear processes which may take place in star interiors, and the growing computational possibilities of fast computers enable us to follow with some confidence the evolution of a star through very long time periods. In particular, we can get some knowledge of the late stages of the evolution. It is in just such very old stars that we can encounter the phenomenon, predicted by the general theory of relativity, of a collapse of a star, which becomes a 'black hole'. In general terms, the evolution of a star is as follows: after the stage of condensation from a cloud of interstellar matter, the star spends a large part of its life in a state in which its total luminosity and surface temperature are approximately constant. After burning up the hydrogen in its core, the star grows larger, more luminous, and redder. In subsequent stages of evolution, a mass loss can occur, and if it is of a very violent character we have a supernova.

If a star is less massive, after a possible mass loss it will then become a dwarf, white star with a density of the order of 10^5-10^{10} g/cm^3, a 'white dwarf'. If in the late stages of evolution of a more massive star, perhaps after some mass loss, the remaining mass of the star does not exceed a few (2 to 3) solar masses, then what remains evolves into a neutron star with a density of $10^{12}-10^{18}$g. This may be a pulsar. If, however, the remaining mass exceeds a few solar masses, the star has to contract to very small dimensions, and finally becomes a black hole. As the general theory of relativity predicts (and was already foreseen by Laplace in 1798), the attraction of such a black hole will be so strong that no particle can escape from its surface. In particular, light quanta—photons—from such a collapsed star could not reach us. If the collapsing star rotates (Kerr, 1963), the resulting black hole—a 'Kerr black hole'—creates very peculiar properties of space-time: energy can be drawn out of it at the expense of its mass. In some cases, the loss of mass (converted into energy) can go up to 30 per cent. This is a very efficient way of changing mass into energy (much more efficient than the usual 0 per cent), cited earlier when discussing quasars and exploding nuclei of galaxies. Possibly in the distant future humanity (or a species into which it will evolve) will also make use of this energy source.

In the last few years much astronomical effort has been concentrated on the problem of discovering black holes somewhere among the stars. It follows from what has been said above that these objects can be recognized by their relatively large masses and very high degree of condensation. The mass of a star can be estimated, as a rule, only if it is a member of a binary system. The degree of condensation can be determined if there is a mass accretion by the more condensed component of such a binary system. In the case of a black hole, and also in the case of a neutron star, the quanta generated in the process of mass accretion will be very energetic, and X-rays will be observed. Up to now, astronomers see only one such candidate for a black hole: the Roentgen component of star Cyg-X-1. Is it really a black hole? It is so by default since, at present, there is no competing interpretation of the observational data.

The lack of direct impact of recent astrophysical discoveries on human consciousness

From the definition of time by Aristotle to the strange properties of space-time around black holes, we have superficially touched upon the many problems in which time and stars tell us something about themselves—and, perhaps, about ourselves. No one can deny the great impact of astrophysical research in many areas of the sciences. The question, however, is how much of this impact reaches the consciousness of the public at large.

In this respect, I am not optimistic. The diffusion from the domain of astronomy into the domain of culture is exiguous. This scientific language of ours, the language of the natural sciences, is hermetic. The tragedy lies in the fact that the achievements of our research, paid for by society, reach the consciousness of this very society only through technological applications, as a rule, secondary with respect to basic research. The situation is worse today than it was at the time of Voltaire and d'Alembert. In the drawing rooms of the Enlightenment, beautiful ladies discussed the problems of gravitational theory. Who cares today about field equations and the uncertainty principle? The Copernican revolution sent shock waves through society. Today even the classical conceptions of Einstein are, at bottom, not assimilated by our culture. Perhaps our contemporary science has grown much too vast and difficult to be assimilated and discussed on the amateur level. And perhaps scientists do not fully see the moral duty falling upon them: to form the consciousness of society.

REFERENCES

BONDI, H. 1962. *The Observatory*, Vol. 82, p. 133.
BURKARD, O. M. 1972. *Gerlands Beitr. Geoph*, Vol. 81, p. 277.
CHANDRASEKHAR, S. 1972. *The Observatory*, Vol. 92, p. 160.
FLANDERN, T. C. VAN. 1975. *Mon. Not. R. astra. Soc.*, Vol. 170, p. 333.
GOT III, J. R.; GUNN, J. E.; SCHRAMM, D. N.; TINSLEY, B. M. 1974. *Aph. J.,* Vol. 194, p. 543.
HOYLE, F.; LYTTLETON, R. A. 1939. *Proc. Camb. Phil. Soc.*, Vol. 35, p. 405.
KERR, R. P. 1963. *Phys. Rev. Letters*, Vol. II, p. 237.
LIN, C. C. 1964. *Aph. J.*, Vol. 140, p. 646.
LINDSAY, J. F.; SRNKA, L. J. 1975. *Nature*, Vol. 257, p. 776.
MCCREA, W. H. 1975. *Nature*, Vol. 255, p. 607.
O'CONNELL, D. J. K. 1975. *Coll. Copernicana III*, 189, Ossolineum.
PEEBLES, P. J. E. 1971. *Physical Cosmology* (Princeton University Press).
PENZIAS, A. A.; WILSON, R. W. 1965. *Aph. J.*, Vol. 142, p. 420.
RUDZKI, M. P. 1901. *Bull. Acad. Sci.* (Cracow), February.
SANDAGE, A. 1975. *Aph. J.*, Vol. 202, p. 563.
TRIMBLE, V. 1975. *Rev. Mod. Phys.* Vol. 47, p. 877.
WEBSTER, A. 1974. *Scientific American*, Vol. 231, p. 26.

FURTHER BIBLIOGRAPHY

BOK, B. J.; BOK, P. F. *The Milky Way*. Cambridge, Mass., Harvard University Press, 1974.
HODGE, P. W. *Concepts of Contemporary Astronomy*. New York, McGraw-Hill, 1974.
OLSSON, I. U. (ed.) *Nobel Symposium 12: Radiocarbon Variations and Absolute Chronology*, New York, John Wiley & Sons, 1970.
PEEBLES, P. J. E. *Physical Cosmology*. Princeton, N. J., Princeton University Press, 1971.
SHORT, N. M. *Planetary Geology*. Englewood Cliffs, N. J., Prentice-Hall, 1975.
UNSOLD, A., *Der neue Kosmos*. West Berlin, Springer-Verlag, 1974.

TIME AND BIOLOGY

Frederic Vester

If we consider the significance of time in its objective (and partly in its subjective) aspect from the point of view of such exact disciplines as mathematics, physics and astrophysics, we must be aware that we are dealing with the non-living world. Here time appears as one of the basic dimensional factors in almost all formulae. But the moment we ask for the meaning of time in life processes, such as biology, psychology, sociology, we obviously move into a completely different realm.

What meaning has time in biology (and for biology)? Is it only the letter t in biological formulae? Is it an objective factor playing its role in the calculation of biological processes? Has its straightforward character of an arrow changed to a cyclic character? And what is its relation to past and future?[1]

Time levels in the biology of man

First of all we should point out that in biology we have five overlapping time levels, each of a different biological significance. The first level is that of the lifetime of radicals of about 1/20,000 of a second. It is the time of enzymatic reactions, of catalysis, of energy transport. One molecule of the enzyme catalase, for instance, splits H_2O_2 into oxygen and water by transforming about 17,000 molecules of H_2O_2 per second, one after the other. The next time level is that of biophysical and physiological reactions: nerve impulses, membrane transport, blood circulation, heartbeat, muscle contraction and relaxation, etc. It lies in the region of a second. The next level is that involving cell division, with digestion and similar processes, taking several hours, and thus about 20,000 seconds. Now, with 20,000 times several hours, we arrive at the time level of (biological) generations of 15 years (puberty). And with again 20,000 times this much, i.e. several hundred thousands of years, we arrive at the time level of evolutionary events.

Time levels in human thought

In our brain, mind and consciousness (where processes take place mostly on the biophysical-physiological level of seconds and where our thoughts about *Time and the Sciences* also take place), we find an analogy for these biological time levels—although they are not as far apart from each other, being on a scale of about a factor of one hundred. Just as the above five levels of the biological world exist simultaneously, so also in our mind, in this informational world, we work simultaneously with different ranges of time: seconds (thoughts, short-term memory), hours (discussions, long-term memory), days (planning in the Stone Age range of the food-collector and hunter), years (planning in the agricultural range of the planter and herdsman) and—which may still have to come—hundreds of years (environmental planning of the future).

There is one particularly important step in the time-levels of our thinking, and it seems to coincide with an evolutionary change in consciousness. A similar change probably took place 6,000 to 10,000 years ago: the Fall of Man, when we left the Stone Age paradise to settle down and become planters and herdsmen. Let us therefore come back to the present: to the special change in our time-span of thought. Such a change will not just stretch time but even add a new dimension to our consideration of reality by transforming one-dimensional time-arrows into multidimensional time-patterns.

Arrows and cycles

A DIFFERENT VIEW OF EVENTS

In addition to the pure stretching of time, e.g. for a new approach to planning, we seem to observe a metamorphosis in the play of events. We realize that, as in astronomy, in the life of stars, we are dealing at the same time with both evolution and cycles. However, the concept of time as a linear advance, time as an arrow, has so far predominated in our thoughts. Unfortunately, this has prevented us from seeing other kinds of events: network events, cyclic events. We have seen some of them as linear cause-effect relations. When we stretch our time-scale, we recognize them as parts of an interwoven pattern. Here the same thing happens as when Stone Age man became a planter. When, by planting a seed, he made provision for the next summer, he became aware of a cyclic event: the period of plant growth.

Today many analyses of human relations take account only of linear cause-effect relations. These, however, are short arrows within a sequence of events that together may constitute a feedback cycle. But the cycle is never seen. The moment we take a greater time-scale into account, the whole cycle may be seen, and our one-way thinking becomes aware of feedback loops. It becomes cybernetic.[2]

If we again stretch our time-scale, we may even recognize this feedback cycle as interwoven in a network of other feedback cycles, and our whole relation to time becomes multidimensional. We see simple cause-effect relations in a completely different light. We begin to think not only in terms of events, but also in terms of system processes. A jump to a longer time-scale may thus expose to view a new holistic system of cycles, nevertheless composed of arrows.

INTERLOCKING SYSTEMS

Both considerations (the single cause-effect step and the overall holistic view of a network) seem necessary for an understanding of our world.

This is already evident in the understanding of cellular processes. On the one hand, molecular biologists have discovered a network of interlocking cycles working by self-regulation via informational feedback. On the other hand, there are the single biochemical steps in these cycles that determine why, under certain circumstances, certain paths are followed and not others. These specific cause-and-effect steps are therefore component steps in larger cyclic processes. However, their direction, their sources, the determination of molecular shapes or the catalytic activation of the 10,000 individual reactions in a human cell are controlled, not by the individual steps, but by the nature of the system's cross-linking.

These findings in the microworld would be of no interest for our study of human society were it not for this peculiar structural principle of interlocking systems in living nature, which create a certain repetition of fundamental features throughout all levels of the biological realm. Each individual ecosystem contains biotopes as subsystems with populations within them; within the populations, the individual living being; therein the organ; in this, the cellular tissue; within this again, the individual cell; then its minute constituents—each fitting into one another like sets of Russian dolls. The cybernetic laws, such as those that govern the production process in the 'factory cell', are obviously the same in the smallest unit as in the largest ecosystems, all together constituting the biosphere as a whole.[3]

By thus searching interlocking systems for their common features, we introduce a cybernetic way of looking at things. We determine what is important in the data, we process this in a manner that is 'programmed by example', based on analogies. This allows comparisons instead of systematical allocations by class and features.[4]

When we consider a sequence of events as a cyclic process, what we consider as cause and what as effect depends on where we enter the cycle. Future and past become interchangeable, at least in respect to their cause-effect relationship. This is nothing new to biology, where the 'future' is programmed in our genes and thus becomes the cause of the present. In a similar way, our thoughts can imagine the future and take it into account in present actions.

It is interesting to note that in our means of communicating we have, on the one hand, our alphabetical writing: a one-dimensional sequence of events. On the other hand, we can communicate by pictures, in which many events are structured as a pattern, and thus presented simultaneously. Even between different ways of writing we find this distinction. From the one-dimensional or digital lettering—a particular form of which is the Morse code—of Occidental writing to the two-dimensional hieroglyphics and the Chinese characters of Oriental writing we take a step from logic to the analogical.

It is not surprising that biology shows a corresponding analogy: cells know either growth (multiplication by division) or function (serving themselves through the system they belong to). For these two purposes, they use different information systems or, to be more accurate, different writings. Growth is accompanied mainly by the activity of the genetic DNA chains (corresponding to the linear or Occidental way of writing), whereas function is accompanied by the enzymatic activity of proteins, the recognition of three-dimensional patterns (corresponding to Oriental or pictorial writing). We see that the distinction between 'growth' and 'function' originates in the first steps of cellular information processing.[5]

If we take the above analogy seriously, then of course what our world needs is neither one nor the other (and not even a mere coexistence of 'causal-logical-thinking' and 'feedback-control-system-thinking') but a genuine symbiosis of the two. In Japan, for instance, with its grafted Western civilisation, both lines of thought exist side by side—an almost schizophrenic situation. The circle lies alongside the straight line. Instead of this, the two could form a new unit in which the circulatory process determined a direction and also changed the course of the straight line into a sort of 'breathing' circular movement, thus stabilizing it.[6]

The above consideration, in my opinion, has strong implications for the understanding of viable systems and as background for considering the changes our civilisation has to go through in its relation to time. This is a question of survival.

The development of systems

GROWTH IN LIMITED AND UNLIMITED SPACE

Theoretically, growth may develop under two conditions: in unlimited space, like the development of a cell culture in an unlimited medium, or in limited space, as with the growth of the human race on earth. Unlike growth in unlimited space, here each phase of increase is connected with significant qualitative changes. An observer from outer space would certainly have noted the sudden increase in population density on this planet over the past 300 years. Above all, he would have noticed a growing density of man's 'systems', such as cities, roads, factories, agriculture, mining and traffic.

These artificial systems of our civilization, with all their dross and excrescences, have grown to a point at which there is scarcely any free space between them—the free space that once compensated for and buffered against our interferences with the biosphere. Such compensation becomes impossible within the present network, the individual sub-systems of which must be understood as part of a new organism; that is, they must change to self-regulation, a new situation demanding new dimensions of control that we have never learned.

WHY HAS UNDERSTANDING NETWORKS BECOME SO IMPORTANT?

The main feature of any system is that it is composed of different cross-linked parts. If something is cross-linked, its individual parts influence each other. Either they influence each other incorrectly and the whole thing breaks down, or they influence each other harmoniously, and the system survives—becoming a sort of new organism in the process. The sooner one recognizes the nature of the interactions in a new system, the sooner one will be ready to take the (prophylactic) measures necessary for its survival.

On the other hand, so dense has the network of our civilization become that any interference with its environment is now much more dangerous than in former times.[7]

In a cross-linked system, generally a new type of consequence can be observed that makes life on this planet more and more difficult: namely, sudden changes in areas in which we seem never to have interfered deliberately. Our influences now have effects that do not end at their apparent target but are seemingly connected by a dense network of invisible threads. Owing to unrecognized feedbacks, they can have the opposite effect to that which was intended—only with different delays.[8]

As long as we fail to comprehend our changed situation and cannot see the cross-linkages, we shall continue to suffer even greater setbacks and will continually have to double our efforts in order to carry on just a little longer in the old way at the expense of increasing energy and material consumption. To comprehend, however, means taking a conscious step towards a larger time-span. Either we extend our time-scale in thought, to grasp the system, or we stick to our previous short-time scale, in which case we should only see heterogeneous arrows but never a system.

We must learn not only that interference at one point will lead to unanticipated reaction via intermediate links, but also that non-linear interactions accelerate many processes in such a way that they can no longer be managed by merely correcting mischief. Population growth, energy use and pollution of the environment offer ample proof of this.

The consideration of longer periods of time, however, leads to the realization that prophylactic thinking obviates the necessity for expensive counter-measures and is more efficient than the isolated treatment of individual symptoms as they occur. Moreover, we have to learn that even the constant repetition of an originally correct decision will never lead to a state of equilibrium. Only through a well-considered dynamic sequence of changing decisions will a system develop to stabilize self-regulation and thus achieve permanent viability.

THE LOGISTIC GROWTH CURVE

Let us look at the evolution curve of mankind. The present age of crisis lies on this curve above the centre of the exponential phase, at a point at which the increasing slope should have changed over into a decreasing slope.

The crisis results from the fact that, obeying a mechanical law of inertia, we continued to follow the exponential slope, instead of beginning the required change of direction into the next stationary phase. It seems that we missed the point of inflexion. Such states of transition with critical turning points have obviously occurred already during the evolution of mankind, e.g. the transition from the era of collector and hunter to the era of planter and herdsman, or from this phase to agrarian and then to industrial society, each time entering a new phase of density.[9]

Transitions from a steady state of lower density to a steady state of higher density are normal procedures. They occur in all living systems: during the growth of cells (halted by contact inhibition), the differentiation of tissue (cessation of growth while taking over function), the metamorphosis of insects and other living systems. Their common feature can be represented by the typical S-curves of logistic growth, the exponential phase always being temporary, the inflexion point always a critical stage.[10]

In all species, from bacterial to humans, there can be observed organizational changes from density threshold to density threshold. The first communication is seen in the intimidation of intruders into one's living space. The next threshold is where communication within the group becomes necessary in the search for food, and so on, to co-operation and mutual aid within the group with corresponding social orders, eventually leading up to the creation of a home and the joint alteration of the environment.

Every density threshold, however, involves the danger of a catastrophe. As soon as formerly isolated environments overlap, the so-called "density-stress" opens up two possibilities. It either makes the species sick, sterile, or aggressive, leading to the destruction of a large number of the population and thus a return to a lower density, or it forces the population to adapt to the new conditions.

The step over a density threshold, therefore, forces change. Each degree of density needs a specific level of communication, a completely new thinking and new behaviour pattern adapted to such a situation. Otherwise, density stress may again reduce the population drastically.[11]

The first signs of density stress can be observed in human civilization. During its growth, the enormous amount of energy and material employed has indeed not led to a highly capable race but to a species that is becoming more and more unstable and less and less capable of sensibly using its technical aids, to a society marked by increasing sickness and declining performance, increasing environmental burden, spreading drug abuse, alcoholism, criminality and the disintegration of family life. Along with the pollution problems from our technology, we should take into account the problem of general stress magnifying the effects of environmental pollutants and distorting our organization and social planning.[12]

The transition from the last stationary phase some thousand years ago (slow linear development in growth, utilization of raw materials and exchange of information) to the present exponential development (population explosion, urbanization, energy production and a torrent of information) seems above all a direct consequence of the fact that the human being no longer just reflected his scientific knowledge, but applied it to his environment and to his own prosperity. Man did this, however, without changing his thinking correspondingly. Although our civilization has thus progressed and expanded, we have still not made the necessary change that is called for at this density stage. We are disregarding the principles of viable systems while retaining an attitude appropriate to an earlier density and to correspondingly fewer cross-linkages.

Worst of all, we believe that all we need to do to overcome the damage our living environment has already suffered is simply to use more and more energy and to produce faster and faster. Our problems cannot be solved by just applying more energy to correcting the mistakes due to our following the myth of economic growth. We are in a vicious circle that will lead to the collapse of the affected system.[13]

A SECOND APPLE FOR ADAM

Without doubt man can change his behaviour, his technology and his economy. As already pointed out, with the beginning of agriculture human consciousness must once have changed suddenly, not only in the time needed for planning, but also in its relation to the environment. The Fall of Man, the eating of the apple from the tree of knowledge, can be understood as the separation of the human race from the ecological community and as the sudden appearance of a new way of looking at its environment.

Perhaps this was necessary to initiate the programme of which the present transition state is a part. But why should we not be able to do so again? A new self-regulating and stable steady state on a higher density level needs nothing more than a change to a new reintegration of man into the biosphere. And this, as already pointed out, is first of all a question of enlarging the time-span of our planning.

Of course planning a hundred years ahead looks absurd but not more absurd than to the primitive hunter, accustomed to think one day in advance, who changed suddenly to a 365 times longer time-scale, by the act of planting his first seed, a seed he actually could have eaten on the same day.

In the earlier state of low density, rules, traditions, laws and other nominal values were consistent over many hundreds of years. Now, with the accelerated growth rate and the development of a more complex network of systems, conditions change at every point of our growth curve so rapidly that even newly established values and norms lose their validity almost as soon as they are introduced—and may become dangerous or even deadly. How can we gain security in a rapidly-changing situation, in the transition state of a logistic S-curve?

If we want to avoid the stress of constant confrontation with the unknown of incompatibility with the constantly changing situation, of loss of significance of fixed norms, there seems to be only one solution: instead of static standards that are very useful during stationary phases, we have to find dynamic norms that are not fixed points in our co-ordinates, but correspond to the slope of our curve; mathematically expressed: the differential of our actual point on the curve. This differential, indicating the angle of the slope, will be constant for a long time—although the point moves away. We could even build into such a dynamic norm the later flattening of the slope in taking the second differential, which takes into account the approach to a new steady state.

Dynamic norms, however, mean that mankind must jettison some existing norms, taboos and habits, such as values of thought, which were established for a static civilization. Others must be understood as dynamic norms relative to former times. In that way, we need not discard old traditions or criticize them as wrong. They can still remain with the new dynamic norms that make them precursors of new temporary 'traditions' by a regular change. Altogether, they will then just be time-bound interpretations of one and the same dynamic norm.

Viable organizations

THE ANALOGICAL APPROACH

The frequent analogies to biological features I have used already may speak for their usefulness. Nevertheless, the use of analogies instead of a purely logical argument has a more substantial point than just attractive metaphor. Analogies are found by comparison. It is difficult to form an analogue by pure abstraction. Its essential is to compare patterns; so again we meet these two ways of communicating: by the alignment of words and by the presentation of patterns. A true analogue therefore must adhere to the principle of looking at things as they are. This 'looking at things' and comparing their patterns is a fundamental feature of the experimental sciences, one of which is biology.

Therefore, if instead of just comparing patterns although describing them in words, we still try to prove their similarity or analogy through one-dimensional logic—which in most cases is impossible—we will never emerge from one-dimensional thinking. We concede an intrinsic truth to the analogical approach.

In this latter analogy lies also the answer to the question: what should we look at if the approach to a viable system cannot be found through one-dimensional logic but rather through multidimensional analogy? Of course, we should look at biological systems. They are the only viable systems we know that belong to an organization that has persisted for about 4,000 million years. The step mankind has to make to survive in this new stage of density is understandable from biology and is rooted in the consideration of time-scales put forward at the beginning of this article.

Man, as a biological being, as an integrated member of the biosphere and organized within it according to the cybernetic rules of surviving systems, can—like any other living organism—survive only with this system and not against it. Profitable symbiosis (as distinct from exploiting parasitism) is the intrinsic organization device of surviving systems. On the other hand, the dependence of the human race on the functioning of the environment and on the whole biosphere will continue in the future, regardless of the level of technology. The more sophisticated this technology, the more sensitive, if not unstable, will become our civilization.[14]

On the one hand therefore, belonging to the biosphere is the basis for man's life. On the other hand, it entails his submission to a rather cruel mechanism: the law of the elimination of disturbing subsystems, which is an irrevocable cybernetic law of the living world. As a consequence of the feedback cycle into which we are built, this guarantees nature's survival by liberating it from subsystems that leave the state of equilibrium, ignore symbiosis, multiply their energy consumption and therefore become a danger for the overall biosphere system. Such subsystems automatically destroy themselves in the process, a process observable in small populations as well as large during evolution. For the most part, it is a question of exponential growth of density stress, but such events may also follow abrupt changes in the environment of the kind we are bringing about at present.[15]

THE BIONICS OF ORGANIZATION

Without doubt, we need a new type of scientific information that can help us to cope better with reality: information on the symbiotic laws between man and the biosphere.

Here reference could be made to the time-binding character of information, which is especially obvious in the genetic programming of living organisms and systems. Man has not only made use of it in the ontogenesis of his organism but also projected these inherent constructional patterns into the outside world many thousands of years ago. Since our appearance as *homo ludens* we have thus reproduced more and more intrinsic features of our biological nature: mechanisms and technical principles like levers, filters, pumps and motors, energy transfer and chemical factories, all of them being structures and technologies long present in a perfect form in nature.

Our replications were of course and still are today less efficient, less elegant, clumsier and more energy-consuming and less appropriate to the biosphere than their originals.

From ancient times, man has copied nature. But his copying has applied only to things easily represented by logic: technologies, structures and, as we saw, writing and printing and thus some part of biological information. There is one package of information still bound in the genes he has never sought to copy, one thing that he has not learned yet from nature: its organization, the way all these technologies are organized and put together in a harmonious system. Man has never cared for the cybernetic laws that govern these technologies. He has not been concerned with the 'before' or 'after' of a product, with its origin or its consequences, its interaction with the environment, but is still mesmerized by the mere act of producing. Here, as in many other fields, he aims at fixed targets, instead of also thinking in terms of circulatory processes. However, both are necessary, as symbolized by our helix.

Cybernetics and reality

THE ALTERNATIVE TO VIABLE SYSTEMS

Cybernetic action requires cybernetic thinking, the first step in which is to learn to think in larger time-spans, so that we can start to think and plan to use interconnections; to plan in feedback cycles. Whereas the extrapolation in single steps in social, cultural and economic forecasting cannot tell planners how self-regulation (and thus viability) can be organized, the analogic way is more promising, finding out how cycles work and then checking for rules and applying them wherever possible. The system in question will thus undergo a change step by step, a flowing equilibrium capable of evolution, the biocybernetic alternative.[16]

Of course, there may be more alternatives to our one-dimensional and almost compulsively neurotic growth-oriented technocratic process than just biocybernetics. The intrinsic secret of bio-processes, however, is not just a theory of self-regulating systems but often the possibility of action corresponding to principles that make the functioning of all terrestrial life possible. In other words, we are not faced with uncertain experimentation but with the reconstruction of working devices already used for millions of years.

On the other hand, given the conditions prevailing on this planet, a system can obviously survive only if the specific cybernetic laws are observed. The question of what principles allow nature to function so rationally is therefore worth studying.

One soon realizes that, in order to be able to maintain a reproductive equilibrium against external influences and to allow development and metamorphosis of the system itself, a living system obeys just a handful of rules, which can be described as the basic laws for viable systems. Ignorance of these or a failure to observe them may have been the main reason for the increasing ecological, sociological, economic and political problems we are facing today. Here is a checklist of such rules:

The predominant feature in the settling of a stable equilibrium is a principle called negative feedback. In a feedback control cycle, an excessive value of the variable to be controlled is noted by a regulator and reduced by means of a control element. In the same way, if the value is too low, it is increased and the variable stabilizes. If we had positive feedback, a high value would lead to further increase until 'explosion', a low value to 'deep freeze'.

Another important rule is the multiple use of everything that is produced. Every process should, if possible, serve more than one purpose.

A third principle is the most economical and effective utilisation of energy, the principle of *jiu jitsu* (the Asian method of self-defense), in which one does not combat force with counterforce but only diverts and controls the force applied by the opponent.

Another point: coexistence of different forms of life to their mutual benefit, called symbiosis. Symbiosis is favoured by diversity within a small space. Large uniform spheres, central energy supplies, pure dormitory towns, monocultures, etc. must therefore manage without the advantages of symbiotic relationships and thus without their stabilizing effect.

Two other important rules are: independence of growth and independence of the product. In any organism, with increasing functions of its cells, the tendency for cell division decreases. Cells with very important functions, like the neurons of the human brain, cease division a few weeks after birth.

The same is true for products. Like growth, specific products are temporary and thus secondary, the function being primary. Thus, electricity companies should not think of themselves as power producers but as energy suppliers, something that can also imply the obligation to reduce power demands or replace energy consumption by alternatives.

The evolution strategy of nature has proved that these and other rules are the internal guides of viable systems. They should therefore apply as well to the system of human civilization, since it is one of the subsystems of the biosphere. Indeed, such a checklist gives a first approximation for doing cybernetics: from the organization of entire spheres of life right down to individual firms, to consumer behaviour, government measures or the design of individual products.[17]

SELF-ORGANIZING BIOCYBERNETICS VERSUS CONTROL ENGINEERING

The establishment of cybernetic rules and the existence of interlocking feedback cycles touches upon the question of freedom. Looking at the beauty of nature, I think we clearly see a much greater freedom in structure and function than linear growth can give us. Linear growth with its one-dimensional character can only be dull repetition without change of quality and pattern.

However, I still see some dangers in cybernetics if we copy incompletely from nature by establishing closed cycles, regulated from outside. Here we would commit the same error in our new organizational analogy as we have committed in many other techniques that remained incomplete imitations of their originals. Living processes take place in a system of open interaction of many cross-linked feedback control cycles whose nominal values depend upon one another. It is this principle of interlocking self-regulation that represents the great difference between biocybernetics and man-made control engineering. Unfortunately, one notes an increase in the number of cybernetic studies and doctoral theses in economics that are oriented to control engineering. Application of this control-engineering principle to a national economy or to environmental problems, however, would lead in the last analysis to an absolute *dirigisme*, a rigid planning. Such central control is, however, completely foreign to living systems. One must therefore assume that it is not viable.

AVOIDANCE POLITICS INSTEAD OF FORECASTING BY EXTRAPOLATION

Many forecasts and analyses utilize simple projection: an extrapolatory approach which, although it may be valid within small sections, neglects phenomena found in larger systems—phase transitions, hysteresis phenomena, collective behaviour, irreversibilities and a whole series of other seemingly non-causal phenomena emerging from cross-link feedbacks.

Predicting and planning therefore should not be aimed at a desirable future but done rather for the analyses of situations that are to be avoided. If such undesirable situations are designed as well as the measures to be taken in order to avoid them, a forecasting policy becomes practicable by applying the principles of surviving systems step by step backwards from the future situation to the present. This evolutionary feedback strategy will not only handle situations far ahead in time, but may improve the mechanism of our present civilization. Predictions by extrapolation, however, will always deviate from the feedback cycle by prolonging the single arrow to a straight line, while in fact the cycle is curved and will finally turn back to its beginning.[18]

Education and network thinking

If we agree that a new behaviour must be developed, then it is not enough to predict a failure and catastrophe. Insight and goodwill alone will not suffice. The change needed is a process of education.

DIFFICULTIES IN TEACHING CYBERNETIC THINKING

In our civilization, thinking and learning are accomplished according to traditions that are laid down arbitrarily or in accordance with ideas that have no relation to the biological function of the human brain. Education and traditional educational institutions thus provide only stored knowledge and not the means to elaborate and utilize this knowledge.

It is therefore very difficult to teach people cybernetic thinking, although this would correspond perfectly with the biology of our brain cells. Essential parts of the human brain structure work indeed by analogy, i.e. by co-ordinating information patterns through their similarity. These parts are not used, or are misused, when knowledge is drummed into our heads without recognizable connections. This makes the recall of information from the brain very difficult, as soon as it is needed in any context other than in which it was originally absorbed.

When we speak of the lack of an awareness of alternatives, we are referring to our way of abstract intellectualized learning, which is reinforced by our verbal thinking—our thinking in the one-dimensional structure of language proceeding essentially along an arrow of time.

Education in accordance with cybernetic principles should establish teaching methods corresponding to the biology of our perceptual functions, allowing the verbal information in our brain to be transformed to pictures that can be processed simultaneously and non-linearly, and so compared to other patterns.

In this manner the present intensive way of thinking (within abstract conceptions) will change automatically to an extensive and operational way of thinking, where the conceptions and associations are seen in connection with their environment.

If a child is asked 'What is red?' he or she would perhaps answer: 'Red is my mouth, red is the sun, red are tomatoes.' To the question: 'What is a chair?' he or she would perhaps say: 'A chair is what you can sit on.' This kind of answer is called operational because the question is answered through examples, comparisons and procedures. This inborn operational thinking is extensive because it includes the correlation between things and the natural feedback function with the environment. It is soon expelled in our schools and replaced by a much narrower learning process seperated from reality: the classification of notions and concepts by other concepts and their arrangement according to class and feature.

At school we do not learn that a chair is something to sit on but that it is a piece of furniture; red is no longer the sun, a tomato or blood, but red becomes a colour. And with this the foundation is laid for a process of impoverishment of thinking, which finally leads, at the university, to those theoretical systems known as disciplines: special branches of knowledge.

A NEW WAY OF 'LEARNING IN CONTEXT
WITH ENVIRONMENT'[19]

The way out of this dilemma must be sought very early: in the way we teach knowledge, one of the main tasks of Unesco. Education based on brain biology is on the move. It is needed not only in order to make learning easier, faster and more substantial. A new biologically meaningful learning (compatible with the biology of the brain) is necessary for the better handling of our environmental problems.

In Germany a study has begun of a new education pattern related to environment. Learning and using what is learned in context with environment (neglected in traditional education) will be the first step.

Biologically meaningful learning means therefore using existing associations and storing information skeletons on which associated knowledge may be accumulated.[20]

DISTRIBUTING KNOWLEDGE VIA FEEDBACK

Scientific innovation does not help to solve the water problem or other urban problems because scientific knowledge is translated into material form without reflection. This, in my opinion, is due to the way innovations are generated and utilized.

As to their generation, science education will need to be provided along the lines of disciplinary biology, based not only on a different (network-like) way of learning, but also on a different way in applying what has been learned.

Before an innovation is incorporated as a new 'member' of the corresponding natural or social ecosystem, its feedback effects should be examined. Antibiotics looked fine at first. So did the 'green revolution'. The energy boom seemed to improve our quality of life. Later, by indirect feedback effects, antibiotics led to the resistance of dangerous bacteria, the 'green revolution' to monocultures and to a dependence on pesticides and soil-destructive methods, and the energy boom made us dependent on more energy and replaced manpower, leading to unemployment as soon as growth rates diminished.

The main emphasis in solving problems of humanity is still placed on technical progress and economic growth, as has been stated here, whereas the problems are basically social and political—and informational. Our brains construct houses and towns out of an abstract thinking—and we complain of their inhumanity. The towns, in order to communicate in a sane way with the people living in them, should correspond to the psychobiological structure of human beings.

As already stated in our study *Urban Systems in Crisis*, all this is the outcome of the fact that not only is learning channelled into subjects, into faculties and into research in special institutes, but also industrialization and technology and all human interference with environment are pursued in a one-dimensional way. In this way the sphere of technical development and its individual fields are isolated from those of information and opinion-making. They are equally isolated from the political sphere, the commercial sphere, that of wildlife conservation, and those of production, marketing, consumption and waste disposal. But decisions which are inclusively discipline-orientated (science), department-orientated (governmental authorities) or branch-orientated (economy) frequently lead to the gravest mistakes, which could easily have been avoided with even a rough knowledge of the network of links connecting these spheres.

Spheres of life and thought that are in reality interwoven are not, however, seen as such. Our conventional terminology (which is oriented to disciplines) and our administration (with its individual departments) present them to us as artificially separated. The links between them have been cut and thus disappear from sight.

Outlook

Our treatment of 'Time and Biology' (both in the sense of 'factor time' and of 'our time') has involved four aspects:

First, the interlocking network of living systems which is difficult to grasp with our one-dimensional logic, proceeding along the arrow-time—in language as in thinking.

Second, the symbiosis of this logical cause-effect thinking with a thinking in feedback cycles where cause and effect and thus past and future become interchangeable: the biocybernetic approach.

Third, a larger time-scale in thinking and planning, encompassing networks and feedback cycles (increasing with the density of human civilization).

Fourth, the difference between linear information (logical) and pattern information (analogical), both used in the regulation of bio-processes and leading to viable systems.

Of course, there are many other biological aspects of time that could not be treated in this article but nevertheless would fit into its scope.

NOTES

1. Some of the thoughts and formulations in this contribution were part of the author's lecture on 'Biology and the Age of Crisis' at the Congress of ICSID held in Kyoto in 1973, and of his study, *Urban Systems in Crisis—Understanding and Planning Human Living Spaces: the Biocybernetic Approach*, Stuttgart, Deutsche Verlags-Anstalt, 1976, submitted to the Unesco Man and the Biosphere Programme.
2. F. Vester, *Das kybernetische Zeitalter—neue Dimensionen des Denkens*, Frankfurt, S. Fischer, 1974.
3. L. Thomas, *The Lives of a Cell*, New York, Viking Press, 1974.
4. For an example of a bionic computer device, see K. Smith, 'A Computer that Learns like the Brain', *New Scientist*, Vol. 43, p. 473.

5. Vester, *Das kybernetische Zeitalter*, op. cit.
6. ibid.
7. Vester, *Urban Systems in Crisis*, op. cit.
8. See P. Ehrlich, *Population, Resources, Environment*, San Francisco, Freeman, 1970; M. Taghi-Farvar and J. P. Milton (eds.), *The Careless Technology*, New York, Stacey, 1973; F. Vester, *Das Überlebensprogramm*, Frankfurt, Fischer-Taschenbuch, 1975.
9. F. Vester, 'Prinzip und Bedeutung kybernetischer Technologien', in J. Wolff (ed.), *Wirtschaftspolitik in der Umweltkrise*, Stuttgart, Deutsche Verlags-Anstalt, 1974.
10. ibid.
11. W. Schäfer, *Der kritische Raum*, Frankfurt, Kleine Senckenberg Reihe No. 4, 1971.
12. F. Vester, *Phänomen Stress*, Stuttgart, Deutsche Verlags-Anstalt, 1976 (also translated into Dutch, Spanish, French, Swedish and Norwegian; Italian edition in preparation). See also H. Schaefer and M. Blohmke: *Sozialmedizin*, Stuttgart, Thieme Verlag, 1972.
13. See D. Meadows, *The Limits to Growth*, New York, Universe Books, 1972; M. Mesarovic and E. Pestel, *Menschheit am Wendepunkt—2. Bericht an den Club of Rome zur Weltlage*, Stuttgart, Deutsche Verlags-Anstalt, 1974; and H. Gruhl, *Ein Planet wird geplündert*, Stuttgart, S. Fischer, 1976.
14. See Ehrlich, op. cit.; Taghi-Farvar, op. cit.; and Vester, *Das Überlebensprogramm*.
15. See Vester, *Phänomen Stress*; and Schaefer, op. cit.
16. Vester, *Urban Systems*, op. cit.
17. F. Vester, 'Cybernetic Control', *Architectural Design*, Vol. 44, I, p. 7; 'Mobilisierung des Undernehmertums zum kybernetischen Denken', 5. *Symp. f. wirtschaftl. u. rechtl. Fragen d. Umweltschutzes*, St. Gallen, 1975; and 'A Superstable Superfactory', *Scala* (English edition), Vol. 5, p. 42.
18. Vester, *Das kybernetische Zeitalter*, op. cit.
19. F. Vester, *Denken, Lernen, Vergessen*, Stuttgart, Deutsche Verlags-Anstalt, 1975 (also translated into Dutch, Italian, Swedish, Spanish and Japanese; French edition in preparation).
20. ibid.

TIME IN PSYCHOLOGY

Paul Fraisse

Time began to be a psychological problem when philosophers ceased to concern themselves solely with the nature of time and its mode of existence and started to think about the way in which the concept of time originates. In its critical phase, from Descartes onwards, philosophy questioned the origin of man's idea of time, fluctuating between the empirical point of view of philosophers such as Locke, Hume and Condillac and idealistic conceptions such as those of Leibnitz or Kant.

All were united, however, in thinking that man's idea of time has its origin in his experience of sequentiality: perception of the sequentiality of change (Hume) or awareness of the sequentiality of his own ideas, through memory (Descartes, Locke).

As is well known Kant, faced with the need to establish a concept of time that would be valid not only for man's sense-experience but also for knowledge, and knowing that the quest for an absolute time, relating either to the physical world or to the self, was fruitless, came to consider time as an ideal concept stemming, not from the subject's experience, but from the very nature of his mental activity.

This epistemological research led to, and was replaced by, what I shall call the first psychological study of time. Nineteenth-century philosophers and psychologists began to study what might, with apologies to Bergson, be called the immediate data of consciousness, where we find not only successive representations—interrelated by virtue of associations and memory—but also a continuity, the quality which Bergson, contrasting it with man's spatialized representation of time, calls duration. This reflective rather than scientific approach finds its present-day sequel in a phenomenological approach to the temporality of consciousness and an existentialist approach that focuses on the personal experience of duration.

Twentieth-century scientific psychology, however, sees all these problems in a new light, through the eyes of the behaviourist revolution. What role does time play in human conduct? This question itself falls into two parts: (a) In what way does time determine our activities? (b) What is man's cognitive behaviour with regard to time? The distinction between these two approaches will become clearer as the problems involved and the present state of our knowledge are defined.

Time as a regulator of our activities

The whole of this section is applicable to animals as well as to man. Change of one kind or another is all around us. The laws governing these changes are exceedingly varied and form the subject-matter of the physical or natural sciences.

I am leaving aside, at this point, the kind of change represented by the evolution of the world and of individual species, which nevertheless manifests itself within each organism in the form of evolution followed by involution during the period between birth and death. I shall say little here about the different types of behaviour corresponding to the various stages in this process of change, which form the basis of psychological study related to age.

PERIODICAL CHANGES

Changes of this type are imposed on us by the fact that we live on this planet earth. Let us consider the two principal ones: the annual and nychthemeral cycles. These determine the activity cycles of the whole of mankind.

The rhythm of the seasons is not merely a meteorological type of cosmic phenomenon. The lives of plants and of many animal species also observe an annual rhythm. Annual rhythms—of the metabolism, for instance—are also found in man, even though it is not yet known whether these variations are endogenous or exogenous in origin, i.e. whether they are successive reactions to different stimuli, without any internal regulation.

The same cannot be said of the diurnal rhythm. The fact that man has always worked by day and slept by night could indeed be regarded as mere adaptation to light and darkness, but we now know that organisms—and especially the human organism—possess what are called circadian regulatory systems. We distinguish today between the time-ordering systems that present a relatively constant rhythm, even when the organism is deprived of all periodic stimuli from the environment, and the connecting processes that link these systems to external periodic changes in factors such as light or temperature. Our organism is a biological clock with its own regulator but capable of being set to a different time, as it were, by what are called *Zeitgeber* or synchronizers.

All our activity is governed by the alternation of periods of activity and of rest throughout a cycle of approximately twenty-four hours. Recent experiments have shown that this rhythm is maintained even when the subject is placed 'outside' time, e.g. in a deep underground cavern where he is isolated from the world and from human contact. Not only the body temperature rhythm but also the activity rhythm remain circadian. Michel Siffre, in 1963, spent fifty-eight days in an underground cavern, during the course of which he had fifty-seven activity-sleep cycles. Allowing for the fact that a slight lengthening of the biological cycle occurs under such conditions, this represents almost perfect accuracy. The biological clock can, however, become inaccurate, as may be seen from other experiments carried out by Siffre. The circadian cycle may become circabidian, so that a subject who is outside time may function for a while on a waking-sleeping cycle of about forty-eight hours. This doubling of the length of the cycle can be seen as evidence that an autonomous and periodic system of regulation exists.

The maintenance of these activity cycles independently of the earth's rotation can be seen without having recourse to such elaborate experiments, and NASA (National Aeronautics and Space Administration) in the United States has come to the conclusion that the best rhythm for its cosmonauts to observe in their activities is the circadian cycle. The importance of this cycle is, in any case, not merely global. Current research indicates that our performance and our learning ability are both subject to a circadian cycle. The connecting process becomes noticeable when one has to alter one's activity cycle in relation to the night-day cycle, e.g. when changing over from day work to night work or making an ocean crossing by air.

The resulting desynchronization can be demonstrated by various biological tests, but also manifests itself in the form of psychological difficulties (tiredness and sleepiness) that persist for several days.

REGULAR SEQUENCES

There are many activities that constitute responses to a succession of signals. One finds in such cases that adaptation to the successive signals occurs in the form of synchronization between the signals and the individual's responses, which thus become temporally regulated. His behaviour no longer consists merely of responses to stimuli; the activity programme now includes a temporal element. Let us take two examples of this in young children. One finds that, at about a year, the child begins to co-operate when he is being dressed. When his mother has put one of his coat-sleeves on, he spontaneously holds out his arm for the other sleeve. He has become adapted to the sequence of operations and shows this by his anticipatory movement. He has become adapted not only to sequentiality but also, in a general way, to the time interval between movements. This can be seen even more clearly in another example, the ability to synchronize foot or hand tapping with regular rhythms. In order to be successful, the response interval has to be the same as the stimulus interval and the two rhythms also have to be in phase. This ordinary kind of behaviour implies the existence of a process of anticipation, since the response has to be triggered before the stimulus in order to be simultaneous with it.

In all these cases, the psychologist speaks of time conditioning. The phenomenon was demonstrated in Pavlov's laboratory in 1907. After a sound and food had been repeatedly coupled together with a ten-minute interval between them, the salivatory conditioned reflex began to appear only towards the end of the ten-minute interval. The same result was found in 1912 with defensive motor reflexes. If an electric shock is applied repeatedly to a dog's paw every five minutes, it is at length found that, instead of being in a state of permanent excitation, the animal remains calm during the interval between shocks, then, towards the end of the interval, it wakes up, shakes its head and raises its paw.

This conditioning is not merely a laboratory curiosity. In the majority of our regular activities a temporal kind of programming enables us to avoid being taken by surprise by foreseeable events.

Our eating behaviour also obeys temporal programming. The programmes vary between cultures and between individuals, but they are sufficiently imperative for hunger to tell us when it is mealtime.

Our knowledge of such conditioning is becoming much more accurate through research carried out on reinforcement programmes, along the lines laid down by Skinner. Using mainly pigeons and rats, Skinner showed that an animal adapts its actions—striking against a target—to reinforcement programmes, i.e. the provision of food. Of these programmes, two have a temporal component:

The FI or fixed-interval programme. When a lever is pressed, food is delivered only if the time elapsed since the previous reinforcement is, say, two minutes. It is found that the animal's responses are concentrated towards the end of the interval, particularly during the last fifteen seconds.

The DRL or differential reinforcement of low-rate programme is more complex. The reinforcement is obtained only if one response follows another at the end of a given interval, say twenty seconds. In this case, at the end of the learning period, the animal's responses are grouped around the twenty-seconds mark, some slightly before it and some slightly after.

It seems worth mentioning that temporal regulation enables us to anticipate events: at the biopsychological level to forestall fatigue or inanition and, in our sensori-motor activities, to anticipate external stimuli.

This temporal regulation has a biological basis and is sensitive to temperature and to the action of drugs. Disappearance of the alpha waveform in an electroencephalogram can be conditioned in the same way as behaviour can be conditioned.

In addition to their adaptive effect, all these mechanisms of temporal regulation provide signposts that help man to orientate himself in time and to estimate duration. We are reminded by hunger that it is midday and by sleepiness that it is time to leave an interesting party. We have at our disposal a biological clock that enables us to take our temporal bearings—provided we have confidence in it. Let us look again at the case of Siffre, who lived outside time for fifty-eight days and had fifty-seven waking-sleeping cycles during that period. If he had trusted his biological clock, he would have gauged the length of his stay almost to a nicety, but he thought he could do better by constructing a calendar based on his estimates of the length of his waking and sleeping periods. This produced a considerable degree of error: when the experiment ended, he thought he had spent ony thirty-three days in the cavern. In another experiment, a subject who had spent 174 days outside time thought that only 86 days had passed.

Cognitive behaviour

Temporal regulation of behaviour is common to both man and animals, but man also has the faculty of being aware of time. This faculty of man's can be seen in all the physical and life sciences, but psychology's rôle, even before its epistemological aspect, is to investigate the working of this awareness.

Our awareness of the universe shows itself in the use of language, which is a system of signs used to establish communication between human beings. These signs possess the characteristic of doing double duty, as both signifier and thing signified. The word I speak as a signifier corresponds to a separate reality, which is the thing signified. The sign is arbitrary and conventional in the relationship it establishes between signifier and signified. This dual aspect plays a vital part in the development of any science that seeks to reduce real phenomena to a rational order, and it is particulary significant in the study of time from the psychological point of view. Man, and man alone, possesses signs whose present signifiers refer not only to things signified that exist in the present but also to those that existed in the past or are planned or conceived for the future.

Language is the instrument that enables man not to be merely subject to the forms of temporal regulation imposed on him by the environment and by his organism. Language enables him to be aware of these changes and, to a certain extent, to master them.

After discussing the way in which man perceives time in the psychological present, we shall consider several aspects of this awareness—aspects in which major scientific advances have been made in the past twenty-five years: the evaluation of duration, the temporal horizon, and the genetic development of the conception of time.

THE PSYCHOLOGICAL PRESENT

Man does not live—any more than animals probably live—in a timeless present, i.e. in a mathematically precise instant separating the past from the future. There is a perceptive present, usually called the psychological present. Just as our eyes take in a given spatial field at each instant, so we have a temporal field which, in a single act of apperception, subsumes within the present the inevitable succession of events. It is possible to understand a sentence because at the moment when we are perceiving the end of it the beginning is still present. After a pause, which is represented in writing by the stop, a new temporal field of view begins with the first words of the new sentence. Within these limits it is not the memory of what has gone that connects it with what is present; there is a direct process of association linking present events. Perception of a rhythm or melody is based on the same phenomenon. A man lacking the ability to experience the field of the present would perceive music in the same way as a blind man perceives area. In fact, he would be unable to perceive the music, but could only reconstruct the succession of notes.

Rhythm and language have been used as tools to evaluate the duration of the psychological present, which cannot be dissociated from the nature of what is being perceived and is dependent on three factors:
First, the interval between two stimuli. If a pendulum were to slow down greatly, there would come a moment when the 'tick-tock' would no longer be perceived. There would be a 'tick', followed later by a 'tock'. Language uses phonemes lasting from 0.15 to 0.35 seconds, and music uses notes of lengths varying from 0.10 to a maximum of 0.90 seconds. It can be said that perceptive continuity is no longer possible beyond two seconds.
Second, the number of stimuli. If a clock strikes three or four, there is no need to count the strokes in order to know the time, but at eleven or twelve it is essential to do so. The field of our psychological present does not exceed 7 ± 2 elements even in the most favourable conditions.
Third, the way in which the stimuli are arranged. We are capable of

perceiving seven letters and also seven syllables; if the syllables are arranged syntactically and semantically, an adult can perceive and accurately repeat a sentence of twenty to twenty-five syllables. A line of poetry rarely exceeds twelve syllables.

When these factors are considered in combination, it is found that few lines of poetry or bars of music last as long as five seconds. In practice, our present instants last from two to five seconds. During that time, we store information in our immediate memory; part of that information is then tucked away in one form or other in our long-term memory and constitutes the raw material that will be used to develop the various forms of our mastery of time.

EVALUATION OF DURATION

When I speak of evaluating or estimating duration, I am not referring to the process which consists in measuring duration by arithmetical calculation. If it is 8 o'clock at time t_1 and 9 o'clock at time t_2, I can say that one hour has passed. The kind of time measurement of which we shall examine the psychological aspect is not a direct evaluation of duration based solely on information furnished by our activity. It is based on the whole range of information that has been stored and can be evoked. To make a comparison, man estimates duration in the same way as a geologist evaluates the age of a sedimentary deposit, i.e. by the number and thickness of the different strata—but he is conscious of the fact that these data can give only an approximate evaluation, since the rate of sedimentary deposition may have varied considerably from one century to another.

Although all the data we make use of are subjective in a sense, since they are data for a subject, they come, in fact, from two different sources: (a) Certain information comes from our environment: noises, sounds, the sequence of days and nights, of seasons, etc. To that extent, therefore, it can be called objective. (b) Certain information is connected with our mental life and relates to given moments in acting, thinking or day-dreaming. This kind may be called subjective. This distinction is useful in stating the problem, but in practice these two types of information are concomitant when they spring from our activities, in which objective data and subjective echoes intermingle.

Psychology has a duty to explain why our estimates of duration, on the basis of information provided by our activity, bear so little relation to measured time and, in particular, why we systematically underestimate or overestimate.

Let us reconsider the general rule: the ability to estimate time is proportional to the amount of stored information available. Rather than cite examples of experimental research, I shall illustrate this by a few phenomena we have all had occasion to observe:

A journey by a new route seems longer than one of equal length by a familiar route. The new route makes us pay attention in order to find our destination, so that we notice far more detail and more occurrences than when following a habitual route.

An interesting activity seems shorter than a boring one. The affective factors involved here should not blind us to the fact that the difference is due to the quantity of information. In an interesting activity, the moments of the activity arrange themselves into a relatively unified whole which may be represented by a task (solving a problem) or a message (reading a detective novel). We are not impulse-recording machines; the activity of the human mind is an organizing activity. To take a simple example, if a subject is given a multiplication sum to do and is also made to write a series of figures for the same length of time, the time taken for the multiplication will seem shorter than the time spent writing out the figures.

Time spent waiting always seems interminable. Waiting is a suspension of present activity until the time when a goal can be reached. An example of this is the child who wants to go and play but has to wait. When we have nothing to do we notice every noise and every thought. Approaching the problem from another angle, what can we do to lessen the length of the waiting period? In a waiting room we can read the magazines that are always made available.

The production-line workers for whom their working time passes most easily are those who escape from the endless repetition of uninteresting tasks into day-dreams. It is now suggested that the tasks should be made more rewarding, by creating longer and more complex units of activity.

The general rule can also be applied to exceptional cases:

1. Situations in which there is sensory privation. Research in this field has increased as a result of the problems raised by life in submarines and atomic shelters. An unexpected result of these experiments is that the duration (whether of minutes, hours or weeks) is always considerably underestimated. Suffice it to recall the case of Siffre, who estimated the length of his stay in cavern as thirty-three days instead of fifty-eight. I would explain this error, which has frequently been observed, by the fact that, in spite of activities needed for survival, life in an underground cavern entails a very low activity level in which there are few events worthy of note.

2. The shortening of time in old age. The older we become, the more quickly the months and years seem to go by, having become, as it were, journeys over familiar roads. In our youth, everything is new, and we store up a great variety of information. The older we grow, the more habitual do events become and the less worthy of notice. In addition,

our ability to memorize diminishes considerably and, all things being equal, the number of items of information stored in a given time becomes smaller.

3. The effect of drugs. There have been many contradictory results obtained in this field. The point must be made that it is difficult to experiment because differences between individuals are considerable and because, for ethical reasons, only small doses can be employed. The criteria used have also been quite variable. One general law can, however, be expressed: the use of stimulants and anti-depressants results in overestimating duration while depressants, tranquillizers and sedative produce the reverse effect.

This last finding can be explained by the fact that stimulants increase activity in general and particularly mental activity, whereas inhibitors reduce it. Several of the hallucinogens, such as hashish or mescalin, produce such a flow of images and ideas that minutes become hours. The effects of LSD are inconsistent. With LSD, as with some other drugs, temporal disorientation frequently occurs and it then becomes impossible to estimate duration.

Such cases can be compared with what occurs in mystical ecstasy or even in psychoses, in which the subject seems to be no longer in contact with reality—but we should remember that reality itself is forever changing, whether within us or around us.

In states such as these the subject can be regarded as absorbed, as it were, in a single mental picture, so that one might say of him what Condillac said of his statue: 'She would never have had more than one moment's experience if the first odoriferous substance had acted steadily on her for the space of an hour, a day or longer.' Reference may be made at this point to Freud's theory that subconscious processes are not affected by change or by time. Almost by definition, there is no connection between these processes and the sequential events of real life. In a state of deep reverie, and when we are on the point of going to sleep, we yield to the pleasure principle and time does not exist, but if the reality principle again comes to the fore, we find ourselves back in the world of sequentiality and duration.

As far as sensory privation, old age and drugs are concerned, we have analysed the effects of these in terms of the production of events that can be memorized. There are certainly, however, other causal chains involved that are more biological in character—or, to put it another way, these processes are certainly mediated by biological or even biochemical agents. It is known, for instance, that drugs have an effect on time conditioning in animals, and that our estimates of time may vary with body temperature. Here we come back to biological forms of temporal regulation, which, we know, are not controlled by the cortex.

There is a basic correction to be made to all this. The foregoing results relate to an initial or primary judgement based solely on a single sequence of representations. In reality, however, we live through several sequences of changes simultaneously, which provide us with contradictory information. Absorbed in my detective novel, I have a kind of impression that time has stopped, but the number of pages read gives me an independent temporal indication. It is possible for me to be bored in a waiting room and yet to be aware, even without a watch, that I have been there only a quarter of an hour.

This multiplicity of information can give rise to a secondary level of judgement that corrects the first (Doob, 1971). This duality is well illustrated by the results observed in subjects under the influence of mescalin. Minutes become hours (primary estimate), but when the subject is asked to evaluate in units of time the duration of his drugged state, he gives a fairly accurate estimate (secondary judgement). This explains the difficulty of carrying out experiments on the valuation of time, because the evaluation criteria can fluctuate from one experiment to another and from one subject to another.

In any case, we often express our estimates of time in the form of an absolute judgement—'it's long' or 'it's short'. The subject is relating his present experience to such standards of duration as the length of a lesson at school or of a meal. The duration of these standards is also influenced by hopes or fears concerning what is to follow the current activity.

THE TEMPORAL HORIZON

Animals live only in the present, responding to stimuli as they occur. These stimuli have managed, at some time in the past, to take on the nature of warning signals that guide the animal's behaviour in relation to the near future (e.g. searching for food or avoiding a blow). Man, on the other hand, has a temporal horizon that is made up of his knowledge of the past and his ability to predict the future.

Leaving aside the methods by which an individual's temporal horizon at a given time can be determined, one can state a few general laws:

1. An individual's temporal horizon depends on the nature of the culture to which he belongs.

The past is a mental construct based on our memories. The articulation of this construct depends on the benchmarks and co-ordinates provided by society in its civil or religious calendar, fairs, feast days, commemorative events, holidays, etc. Our past is arranged around memorable events that occur in the communities to which we belong (family, town, country, etc.).

The more developed the culture is, the greater the individual's participation in that culture and the richer his past.

It should be recalled, however, that the recording and dating of events that have been experienced are also dependent on biological conditions. By studying patients suffering from the Korsakow syndrome, we know that an affliction of the mamillary bodies and of the vegetative nervous centres at the base of the brain can weaken memory fixation. One such patient was capable of agreeing that, according to the calendar, he was 50 years old, but there were so many lacunae in his past that subjectively he considered himself to be 32 or 33.

As far as our view of the future is concerned, it is primarily a projection of our past into the future. It is very significant that in children of 4 to 10 the ability to evoke the past is entirely symmetrical with their anticipation of the future. When the child is capable of using and understanding the adverb 'yesterday', he is also capable of using 'tomorrow'; later he will build up the idea of last week and next week, last year and next year. As for serious mental defectives, their temporal horizon will never extend beyond ten days or so. As a generalization, we can say that the uprooting of an immigrant labourer from his culture impoverishes his past, and his future is limited to his work cycles—his working week and his week-end, his working year and his holidays. Let us compare this temporal horizon with that of a farmer. For him, his past is the past of his farm, his lands and his village; his future is that of his ambitions for enlarging or modernizing his holding and of setting his children up in life.

Resistance to frustration is also linked to different cultural levels. The future is what is desired and not yet possessed; this results in frustration. Working for the future on a long-term basis implies acceptance of this frustration, as when one spends years working hard for an examination in order to attain a certain social position. To give another example: in Africa, groups of people who had attended school for varying periods were asked the question: 'Would you prefer to be given 5 pounds sterling immediately or 50 pounds in one year's time?' Thirty-six per cent of those in the poorly educated groups but only 18 per cent of those who had gone on to higher education preferred an immediate gift (Doob, 1960).

2. The temporal horizon depends on the age of the individual.

From adolescence onwards, everyone positions himself more or less consciously in relation to the length of his past and his expectation of life. Hence the weight of the past increases with age while the scope of future plans diminishes. And yet the old go on making plans for the future right up to their deaths, as long as they continue to have a place in social life and maintain some sort of activity in it. The future loses all meaning only for those old people who are helpless, or for the inmates of old people's homes who no longer have the opportunity to make future plans.

GENETIC DEVELOPMENT OF THE CONCEPTION OF TIME

An organism that experienced only a single series of changes could have an idea of duration—of an interval between present and past or future—but, even possessed of cognitive ability, it would have no ordered conception of time. A child of 3 or 4 is more or less in this situation.

As has already been stated in connection with the evaluation of time, we experience several sequences of changes simultaneously. Thus clock time exists side by side with our activity time. Measuring time, for example, involves being able to relate one series of changes to another used as a standard because it represents change occurring at a constant speed.

The measurement of time has its own scientific history. It also poses a psychological problem that becomes very evident when one takes a genetic point of view, as Piaget did (1946). A young child of under 5 is unable to compare accurately the duration of two contemporaneous events. When an adult witnesses or takes part in two contemporaneous series of changes, he is able accurately to assess the relative duration of the two events by comparing the order in which they start with the order in which they finish. We are able in this way to judge the relative lengths of time during which two moving bodies are in motion by comparing the sequence of departure and arrival, without taking into account distances covered or velocities. Perceiving a sequence of events, followed by durations, followed by another sequence, we are capable of distinguishing the pattern in which the successive durations and sequences of events fit together. This faculty of co-ordination is a sign that we possess a sufficiently abstract conception of time to enable us to carry out any logical operation involving time.

Why is it that a young child cannot do this? He comes up against two difficulties. The first concerns perceptive interpretation of the facts. Sequentiality in relation to two contemporaneous events is not easily dissociated from spatial data. If, for example, two runners set off together and stop together, did the one who has run further stop at the same time or later than the other? The child is uncertain and makes mistakes when the data are contradictory. As for duration, it is estimated by a child (as it would be by an adult) by reference to the changes perceived, giving quantitative data in the form of work accomplished, distance covered and when kinematic data are involved, velocity. When the duration is proportional to these data, no mistakes are made, but mistakes appear as soon as there is no longer a directly proportional relationship and, even more so, when the relationship is inversely proportional. A child will readily affirm that 'more quickly' implies 'more time', but he may equally well centre his attention on the slower, and say that 'more slowly' implies 'more time'. When comparing two durations, a young child fastens on a single datum, which leads him to make estimates that are either true or false but that he cannot explain.

A child's progress is achieved by 'de-centering', so that he becomes more and more capable of taking into account all the data relating to a given situation. He dissociates, for example, the sequence of starts and finishes from spatial data, and evaluates duration by taking into account several factors together: distance, speed, work accomplished, changes noted, etc. This de-centering leads the child to compare all the items of information experience provides him with, and to discover a balanced structure linking them. He then becomes able to dissociate time from its perceptual content, and achieves the abstract conception of time which is that of philosophers and scientists.

Impact on society and culture

The most important result of scientific discoveries in the psychological study of time is a better understanding of the difficulties man experiences in relation to time, at various stages in life and in both normal and pathological states. For this reason, it would be useful for knowledge of this kind to be widely disseminated during secondary education and as part of higher education in the humanities and in medicine. There are two areas, however, where the dissemination of such information is particularly needed.

THE TRAINING OF PRIMARY AND LOWER-SECONDARY LEVEL TEACHERS

A child is slow to acquire a conception of time, but such a conception is none the less essential if he is to understand the elementary problems involved in time measurement, the practical consequences of time-zones in geography, or the chronological pattern presented by differing series of historical events. Even when, around the age of 7 or 8, a child becomes capable of dovetailing sequences and durations in practice, his conception of time remains closely linked to the content of his experience. Not until he reaches adolescence does he understand that time is an abstract concept, and that the world order cannot be changed by resetting clocks, for example.

THE QUALITY OF LIFE

If one accepts that the extension of man's temporal horizons helps him to achieve his full human stature, it would seem that the following points should be given due weight in the life of society:

The way in which work is organized

Everyone has to work, and it is clearly desirable that this work should be no more irksome than necessary. The subjective duration of work is one aspect in its qualitative appraisal. If work is not to seem unbearably long, it is necessary: (a) to provide increased motivation for the activity itself and not merely in terms of rewards such as pay; (b) to create activity sequences that are sufficiently structured to give a sense of continuity.

Attitude towards the future

In so far as this depends on the cultural level of individuals and communities, school and society should be organized in such a way as to maximize participation by all in the life of the communities to which they belong (family, municipality and nation). Hence the importance of the teaching of history, of organized activity, of festivals, etc. This form of culture directly affects man's ability to make plans, and to accept the inevitable present sacrifices involved in any planning activity in personal or national affairs.

Retirement and old age

Although it is inevitable that men become less active with age and even retire from work, it is none the less vital that retirement should not be seen as the end of life—hence the importance of social measures enabling retired people to maintain their activity and to remain in charge of their own lives for as long as possible.

BIBLIOGRAPHY

COTTLE, T. J.; KLINEBERG, S. L. *The Present of Things Future.* New York, The Free Press, Macmillan Co., 1974.
DOOB, L. W. *Patterning of Time.* New Haven and London, Yale University Press, 1971.
FERREIRO, E. *Les Relations Temporelles dans le Langage de l'Enfant.* Geneva and Paris, Droz, 1971.
FRAISSE, P. *Psychologie du Temps.* Paris, Presses Universitaires de France, 2nd ed. 1967.
FRASER, J. T. (ed.) *The Voices of Time.* New York, George Braziller, 1966.
MEERLOO, A. M. *The Two Faces of Man: Two Studies on the Sense of Time and on Ambivalence.* New York, International University Press, 1954.
OENSTEIN, R. E. *On the Experience of Time.* Baltimore, Penguin Books, 1969.
PIAGET, J. *Le Développement de la Notion de Temps chez l'Enfant.* Paris, Presses Universitaires de France, 1949.
ZELKIND, I.; SPRING, J. *Time Research. 1172 Studies.* Metucken, N. J., The Scarecrow Press, 1974.

TIME AND THE FUTURE SENSE

John McHale

The concept of time is woven into our earliest social and cultural beginnings. All peoples have an awareness of its passing, and have explored possibilities of predicting and controlling events in future time.

In the larger sense, though we speak of past, present and future as a continuous succession in time, these divisions are neither wholly linear nor successive. Time, both as collective history and as subjectively experienced, seems a more fluid continuum. What is past, what is now present and what is to come are all complexly interrelated. This may be expressed succinctly as follows:

The future of the past is in the future.
The future of the present is in the past.
The future of the future is in the present.

What was a future becomes, in time, present and past. The future of both present and past also remain open to change. As has been noted, 'The peculiarities of (this) permit us to make the somewhat startling, yet perfectly correct, statements that tomorrow today will be yesterday and that yesterday today was tomorrow.'[1]

There is in the social sense of time the assumption that the past is unalterable—and the future unknowable. But both nations and individuals revise and reinterpret their histories. Memory reshapes our biographic past in the same manner that our selective perception screens the present and the future. We tend to rearrange and reselect events and impressions of the past. Their realignment in the present alters both our past and future. In terms of such plasticity, the past and the future may be regarded as more under our control than the tenuous present.

The need to order this fluidity of subjective time for collective purposes has given rise to a variety of cultural constructs and social institutions. In cultures, the element of myth plays a strong role in time-binding. As Eliade suggests, 'The myth relates a sacred history that is a primordial event that took place at the beginning of time'.[2] Most myths are thus concerned with the past, with the origin of things. As sacred history, they are immutable, unchanging and, as such, unchangeable. Their ceremonial re-enactment establishes social periodicities that give an endurable structure to time. Though referring mainly to the past, both sacred and secular rituals bind the present to the past—to serve the future. Rites of passage re-enact the myth of origin, but the rebirth is to an altered future state. Fertility ceremonies commemorate past periodicities as propitiations for future continuance. Many utopias are mythic returns to the idealized beginnings of a perfected form.

The social ordering of time regulates our collective activities. Whilst co-ordinating the sense of present and past time, it establishes predictability and shared anticipations with regard to future time:

> Expectation, intention, anticipation, premonition and presentiment, all these have a forward reference in time . . . Implicit in all our actions are plans, however vague and inarticulate, for the future, and sometimes, as in savings or investment, this planning is deliberate. As we ascend the evolutionary scale, the temporal horizon becomes more and more extended . . . In man the horizon may reach beyond his own brief existence: from infancy onwards there is a growing capacity to relate what is happening at the moment to events foreshadowed in the more and more distant future.[3]

Individuals, cultures and societies may be modally oriented towards the past, the present or the future. This major orientation has a strong value component that influences the direction of personal and collective actions. Apart from the value placed on time itself, on its duration or on recording its periodicities, the prevailing temporal perspective will determine to a considerable degree how time is invested. Time, space, energy and other resources may be allocated, with varying emphasis, to the service of past traditions, to present needs or future prospectives.

Despite the modal orientation of a society, the centrality of the future in human affairs is attested to by the great variety of social roles and institutions accorded to the prediction and control of future events. These range historically through the oracular clairvoyance of the shaman, the medium and fortune-teller to the larger social function of the priest, king, prophet or political leader and to the more explicit professionalism of today's planners and forecasters. The ascendancy of science over religion, in our recent past, is partially based on claims for its greater predictive capacities, derived from direct observation and measurement of physical phenomena rather than inward contemplation.

Our current outlook on the future is relatively new in human experience. The ascendancy of the future as an object for rational knowledge and exploration, though recent, has its origins in a series of complex conceptual changes and their accompanying shifts in social and other paradigms.

The future past

Most previous societies have operated with quite different models of the past, present and future of humanity, of societies and of the universe.

For some, the future was essentially a continuation of the past beyond an unchanging and unchangeable present. Their view of any future state was limited by their time perspective, by short-term survival constraints in the present and by lack of a basis for conceptualizing radically different future states.

The closer realities of life were birth and death, which bounded the known and familiar; the remote future belonged with the unknowable. Magic, religion and art mediated with the unknown and provided symbols for its propitiation that clothed its awesomeness in quasi-familiar and tangible forms. The future as life-after-death of the individual was a move into the past to join those who had gone before.

Many societies, particularly in the East, operated with a cyclical model of future change, of recurring cycles of individual birth, death and rebirth, whose predestined sequences moved towards the unknowable—a nirvana state of tensionless merging with the universe. Others shared a circular model with an ultimate earthlike paradise—a place of easeful plenty, hero's Valhalla or sybaritic dream. In both, the eternal return was to the source of all origins.

The future present

Our contemporary view of an unfolding linear future, which is relatively unique and recent in the historical sense, is a specific social and cultural development. It embodies within it the idea of progress—both material, in terms of the improvement of human welfare in the present or near future, and metaphysical, in terms of the perfectibility of human institutions and, to a degree, of the perfectibility of the human condition. As a 'world-view', it crystallizes typically in our own period, which is marked by an acute sense of historical time and of the relative nature of temporal, spatial and other phenomena that were previously considered to be more absolutely fixed and determined. It is a period also in which we now extend our knowledge and measurement of past time almost in due ratio to the degree that we project activities and concerns into future time.

This paradigm of radically different future states, dependent upon varying measures of human intervention and control, has multiple origins. It has developed over time from a wide variety of contributory sources. These may be grouped somewhat arbitrarily, though they form part of a larger pattern of interrelated conceptual revolutions whose elucidation goes beyond our present bounds.

The idea of social progress

For the greater part of human history, this idea was not commonly held. The ancient societies of Egypt, Persia, India and the East were oriented somewhat differently. There was a long unchanging fixity of world-views, and a rigidity about the goals and purposes of such societies, which did not include any doctrine of temporal progress toward some altered social and physical conditions of their populations.

Our current views on linear progress and improvement in the human condition appear to derive from two closely associated sources—the Greek and the Middle Eastern prophetic traditions. These, of course, emerged from, and were influenced by, earlier traditions in other parts of the world.

The Greeks initiated a transformation from prevailing religio-mythical views to one that was more intellectual and rational in its vision of the relationship of the human to the universe. Having reduced the god myths to more human dimensions, the Greek thinkers have now become mythicized into intellectual hero figures whose sacred texts, such as Plato's *Republic*, are still invoked to legitimize our major political and intellectual institutions.

The Middle East traditions, based on monotheism, were more directly prophetic with regard to the world to come. Judaism carries forward the idea of a rational covenant with a single god figure through which its people may reach the 'promised land'.

The utterances of the Judaic and Islamic prophets are a continuous, repetitive exhortation to follow the law, lest both society and the individual fail to enter both the earthly and divine utopias of the covenant.

The importance of this tradition with reference to change and progress is its emphasis on linearity and on the conscious human act as a factor for measurable social betterment:

... for the first time the prophets placed a value on history, succeeded in transcending the vision of the cycle (the conception that ensures all things will be repeated forever), and discovered a one-way time ... Historical facts thus became 'situations' of man with respect to God, and as such they acquired a religious value that nothing previously had been able to confer upon them.[4]

Magic was displaced in favour of individual ethical responsibility as the change force in society, and religion was turned from ecstatic propitiatory ritual into a form of progressive social criticism.

The idea of personal redemption in the Christian tradition took this process a stage further. Individual and social pasts become mutable. All people are chosen and redemption extended to all. History, change and progress are also reborn in this process—and made universal.

Though conversion and redemption are primarily religious ideas, they carry with them the notion of intervention and control over past and future, and the possibility of ameliorative social progress through the conscious act:

Change, in short, could be the result of individual purposive action, pitted against the forces of present reality. The transfer of the power to change from God to man comes with the concept of possible breaks in the former smooth continuity of time.[5]

Within this continuing tradition there were further messianic and apocalyptic movements, through which visionaries sought to prepare the people for 'the new society'—to come about through renewed socio-ethical conduct. This immanent feeling of abrupt change to some idealized future state is particularly found in the millenarian visions that presaged the end of the Middle Ages.

Two key bench-marks more specifically related to the materiality of the idea of progress are, of course, the Renaissance and the Reformation. The first gave impetus to the growth of scientific control over the environment by the systematic derivation of logical principles from the direct observation and measurement of natural processes. Predictions regarding physical behaviours could be verified by reports of experimental evidence. It is significant also that, though the Renaissance marked the beginnings of new modes of predictive control of the future, this was duly accompanied by a rediscovery and re-evalution of the past in its preoccupation with the Graeco-Roman heritage.

The Reformation, ostensibly a spiritual rebirth, buttressed the change towards materiality and more conscious control over social futures. The Protestant ethic emphasized gainful activity albeit in the service of God but towards the accumulation of wealth as a sign of His favour. Waste of time itself became a sin. Life was to be ordered economically and organized in the progressive attainment of materially evident grace. Such redemptive progress, however, was not to be measured by the display of wealth but by its 'abstract' accumulation—not only of spiritual but of material currency—by the growth of mercantile ventures, corporate enterprise and, significantly, by the deferment of immediate gratification for future gain.

These movements towards a new future orientation were increasingly accompanied by individual contributions now more widely diffused by the invention of printing. More's *Utopia* and Bacon's *Novum Organum*, with its ringing challenge, 'The greatest single idea in the whole history of mankind is the vista of possibilities which opens before us', lead forward to Descartes' first title for his *Discourse on Method*, 'The Project of a Universal Science which can Elevate our Nature to the Highest Degree of Perfection'. Vico's 'New Science' introduced the key idea of evolutionary historical progression by positing that human progress was not linear but a spiral in which every turn is higher and more advanced than the last. The full tide of reason in the service of man was expressed by the eighteenth-century philosophers. Voltaire, Rousseau, Diderot, Montesquieu, Condorcet, Turgot, Godwin and Paine are an extraordinary roll-call of those whose prime vocation was to speculate rationally on the future of the human condition.

Referring to Mercier's *The Year 2440*, published in 1770 and the prototype of many later utopias, Bury notes: 'As the motto of his prophetic vision Mercier takes the saying of Leibniz that "the present is pregnant of the future".'[6]

Though the writings of the eighteenth-century philosophers are usually considered the formal origin of 'the future sense' we may, however, underline, at this point, that the Enlightenment was really its more visible flowering. Material progress and the possibility of social betterment had become more physically evident by that period. This in turn had generated wider expectations amongst more people. A new social ordering of the benefits of progress were being demanded. These urgings and pressures culminated, of course, in the French Revolution, an event that not only shattered the old order of things but typically sought, in its latter phases, to refashion time itself in a new calendar.

In terms of developmental coincidence, the 'other revolution' of industrialization, also under way in this period, was based more directly on the manipulation, control and measurement of time and energy for the mass production of goods. The possibility of a materially better future for all people was being given physical reality.

The waves of social, economic and political change incident on the demise of the old order, and the growth of industrial society, gave rise to a new generation of prophets—Saint-Simon, Fourier, Comte and Marx. Their acute awareness both of social disruption and its potential for reordering society produced a spate of secular and quasi-sacred views of the future that are still a central force today.

Though much nineteenth-century thought is characterized by material optimism regarding the future, there is a beginning estrangement from modern society that also persists into our own period. It is particularly marked, for example, in the social sciences. Max Weber's phrase, 'The fate of our times is characterized by rationalization, intellectualization and, above all, by the disenchantment of the world', sums up the feeling of malaise and nostalgia. It is more evident in the romantic revival of mediaevalism, in the pre-Raphaelite dream, in Ruskin, Morris and others, and in the advocacy of a return to the simplicities and old stabilities in the work of monastic utopians such as Fourier. Modern man may find solace only through allegiance to the larger solidarities. The reduction of alienation and anomie, the way to social health, lies with emphasis on past normalcies, on the group and on conformity to collective norms.

This ambivalent attitude towards the future continues into our own period. In the West it is evident in the erosion of the idea of the inevitability of material progress through science and technology, in the rise of the counterculture, and even within futures studies themselves with such models as *The Limits to Growth* and others.

Paradoxically, of course, just at the point of intellectual disenchantment with the idea of material progress in the developed world, the lesser developed countries have vigorously embraced many of its aspects. The future is to be created in the present, and for many the prophet of that future is Marx, with his particular view of the historical inevitability of certain modes of social and temporal progression.

The idea of future social and material progress bifurcates therefore in our time. One line abandons linearity and inevitability, and the other, in many ways, seeks to develop it further.

Contributory changes via science and technology

In their pervasive impacts on the future sense, the changes introduced by science and technology may be said to originate with the conceptual revolutions in science around the early 1600s. Technical improvements in the measurement and recording of time lie further back. Hour glasses, water clocks and sundials had existed for a long time:

> ... by the thirteenth century there are definite records of mechanical clocks and by 1370 a well designed 'modern' clock had been built by Heinrich von Wyck at Paris. Meanwhile, bell towers had come into existence ... and the regular striking of the hours brought a new regularity into the life of the workman and merchant. The bells of the clock tower almost defined urban existence. Time-keeping passed into time-serving and time-accounting and rationing. As this took place, Eternity ceased gradually to serve as the measure and focus of human actions.[7]

This change in temporality sharpened during the Renaissance with the notion of the extension of rationally predictive control over physical events inherent in the scientific method. Both the date and the discipline mark the end of one kind of culture of long-established dominance, and the beginning of a new and unprecedented form. Since that time, virtually every notion and cherished belief about the nature of the physical universe, of society and the human place and function within it have been slowly eroded, modified and, in many cases, swept away.

At first the changes visibly wrought in society by new scientific discoveries were relatively slow. They gained social momentum only when such principles began to be applied in developing industrial technologies in the late eighteenth and nineteenth centuries.

Their intellectual momentum has been noted in reviewing the idea of progress, but there was a specific set of changes in conceptual time perspectives introduced by other thinkers in this particular period. Although they were primarily concerned with the extension of the past, some of them profoundly affected the sense of the future.

The elaboration of the geological fossil record, and developments in palaeontology, paralleled Darwin's theory regarding the origin of species, to give an expansion of evolutionary time that strikingly altered the whole temporal perspective. Comparing this to Galileo's expansion of our ideas about the scale of the universe, Toulmin suggests:

It is a transformation whose significance has been much less widely recognised—perhaps because it is still continuing and its effects are not yet complete. I shall call it 'the discovery of time'. Already—in a little more than 200 years—this has multiplied the temporal span within which we organize our knowledge of the past more than a million times; and we may count such intellectual influences as Darwin's theory of natural selection amongst its byproducts.[8]

The scientific idea of evolution itself, as an 'improving mechanism', appeared to buttress the feeling that physical and social laws of development were congruent—and more subject to predictive speculation. The calculations of Malthus suggested limits to human expansion, but these tended to be swept aside by more positive claims for material plenty.

Marx and Freud emphasized, in their separate ways, the new-found future plasticity of the human condition. The former sought the levers for future change in evolutionary historical laws leading to revolution; the latter located them internally—by 'reliving' his past in the present, the individual could find future freedom.

In the arts, Romanticism gradually inverted. Orientation towards the past gave way to Wellsian speculation about the future. The modern movement in architecture and design, though born in revolt against the machine, slowly began to interpret its changes, reflect its images and control its protean qualities through more rational and functional forms.

The invention of photography significantly fixed moments in time with a realism that freed the plastic vision for other pursuits. In the beginning of the twentieth century, one key group of artists adopted the label of the Futurists, who embraced speed, technology and violent change in their manifesto; Cubism presented the past, present and future of the pictorial object for simultaneous viewing.

More rapid transport and communications—the railroad, telegraph, automobile and eventually the telephone and the aeroplane—enhanced both physical and psychological mobility, decreasing the temporal constraints of distance. Life expectancy itself improved as many formerly killing diseases were brought under control, and the future possibilities of a longer life-span for more people became evident.

As we entered the twentieth century, the human condition seemed to be capable of unprecedented expansion—in knowledge of the past, control over the present and optimism regarding the future.

But the most abrupt and fundamental of these transitions occurred when experimental science in the late nineteenth century began to extend its measurable range into the invisible subsensorial world of atomic, molecular and radiation phenomena.[9] A new state of intangibility and indeterminacy was introduced into human affairs. From being a relatively contained, fixed, determinate and 'rationally' apprehendable Newtonian universe to the generally educated person, the whole order of reality began to shift its outlines, become ambiguous and infused with relationships that were neither visibly nor logically apparent before. Alice not only voyaged into Wonderland, but went through the Looking Glass.

In this new Einsteinian universe, future ends and present conditions became, like waves and particles, mutually complementary and identical. Many of its dilemmas are still emerging in our own period, and now not only profoundly affect our sense of the future but also raise questions about the future survival of human society. The ambiguities and uncertainties which began to retard the vision of expansive future progress took awesome form in the mushroom clouds over Hiroshima and Nagasaki.

The changes in change

Marking 'the onrush' of the future in our century have been the more visible effects of changes themselves. They are more directly perceived within the average lifetime. What was to be in the future appears to enter the present more and more swiftly.

The underlying pattern has altered not only in the number of changes occurring but in the quality and degree of their relationships—with increased interpenetration, feedback and interdependence of one set of events upon another. Where societies have previously dealth with relatively separate change factors, within local and limited circumstances, these are now 'global' in their temporal and spatial dimensions. They are no longer isolatable sequences of events separable in time, in numbers of people affected, and in the social and physical processes perturbed.

Certain specific aspects of our contemporary sense of the future have, therefore, been engendered by our exposure, in a very brief period, to what may be the greatest acceleration of change in human history. In just over one hundred years—roughly three generations—successive waves of scientific, social, economic, political and technological changes have crested one upon the other. An explosive increase in human numbers has been accompanied by a concomitant growth in human activities impacting upon the physical environment, and by increased transformation of the planet to human purposes. The range and scale of our collective actions may potentially affect the balance of life on earth and reach into almost all aspects of individual and social living. This temporal compression of processes and events may be characterised as an 'evolutionary' magnitude of change occurring within a historical, rather than geological, time frame.

Rather than merely being a developmental unfolding of the future sense from the past, our particular circumstances now constitute a major change in that sense and in the sense of time itself.

Accompanying the acceleration of change, there have been a perceptible 'time dissonance' and a sense of discontinuity—for example, the disruption of generational time linkages in the transmission of norms and traditions as a variety of 'generation gaps' arise, with their accompanying conflicts in values and attitudes; changes in the life cycle with the emergence of 'new' time-span allocations such as prolonged adolescence, and longer retirement as life expectancy advances; cultural changes via the growth of the mass media, leading to discontinuous changes in expectations and life styles due to the wide diffusion of rapidly changing images and models of human conduct.

The tempo and nature of work have changed towards more precise and tighter organization and more detailed time budgeting. Many industrial work processes continue through night and day; machine time has no biological clock.

Both work and leisure have more visible temporal 'costs' as their demands become co-equal in many societies. Time is both more scarce and more abundant in the sense of being more individually disposable.

The speed of communications and transport outruns the earth's time zones, and requires organizational adaptation to 'uniform time' to carry out many functions and services that are increasingly global.

In developing societies, the adjustment to new temporal organization can be one of the most difficult areas of adaptation to change, clashing with traditional socio-cultural time values and disrupting the whole temporal context of the society.

The 'time cushion' for decisions in both private and public life has decreased as the number of impinging events appears to increase. Shortening of the period between the occurrence of a problem or issue, its news reporting and demands for its resolution tend to thrust public decision-makers into daily crisis-management.

The psychology of decision-making under various time constraints is still largely unexplored. For example, the time constraints on political decisions in societies with elected administrations, operating on short electoral mandates, are quite different from those in other kinds of administrations in which decisions can be taken to commit the society to plans for five, ten or twenty years, without undue reaction from the public and the press. Where political office-holding may be dependent on the success or failure of decisions over a few years, the tendency may be to cast all policies in terms of their shortest-range payoffs, with a consequent neglect of those that extend beyond the period of office. This may be one of the political problems of the present, i.e. how to make short-range tactical decisions with reference to issues whose consequences and effects may not be felt in less than ten or twenty years.

The arts of the period reflect, in technique and imagery, many of the above 'dissonances'. Cinema and television, with their synoptic presentations of interpenetrating temporal events, give a simultaneity to past, present and future, with capacities for 'freezing', instant replay, and other transformations of time that are unprecedented. In music, painting and literature the dislocation and re-juxtaposition of previously linear images and intervals have been major preoccupations. Former unities of time and space have given way to a more plastic and co-extensive temporal context.

The efflorescence and popularity of science fiction may be mentioned here as an important cultural indicator. Having gone far from the elaboration of utopias and dystopias in the Huxleyan and Orwellian tradition, and from more simplistic technological extrapolations, science fiction is now an established literary genre which plays an important role in forming our public and private images of the future.

As the resultant feeling of imminence and urgency diffuses through the wider populations, through increasingly swift mass communications, the collective sense of the future has again been subtly changed and polarized. For many, anxiety and apprehension have replaced optimistic anticipation as the tenor of individual lives is disturbed, both by the personal impacts of change and by the wider diffusion of information regarding critical events. Threats of war, possibilities of environmental catastrophe, terrorism, economic instabilities and civil disorders are common everyday news. But where these were previously local and contained, they are now writ large on the world scale. For others, though the idea of inevitable progress to a better future has been tempered by the negative aspects of unintended consequences of rapid change, the future remains a challenge—a more immediate area for interest and speculation than ever before. In both cases, the sense of the future has been sharpened and more closely woven into the temporal fabric of everyday life.

This new awareness of change, and of the longer range and increased scale of the consequences of human actions, is part of a larger shift in human consciousness and conceptuality. There is a change in paradigm, 'world-view', or the governing set of ideas, which begins to alter interpersonal relations, those of the individual to society and of human interactions with the 'planetary habit'. It is both temporal and spatial. It is both inwardly oriented, from intense re-evaluation of human purposes, towards the unravelling of the micro-life code at the molecular level; and outwardly, towards the successful maintenance of human life beyond the earth's atmosphere and under its oceans, towards the distant monitoring of other worlds and galaxies.

Predicting, forecasting and planning

The question of the predictive record is a difficult one. Detailed discussion of specific successes and failures in the recent past would be lengthy, and would furnish little that could be generalized across different fields. There is nothing comparable to the long-range predictive capacity of astronomy, which rests on a cumulative record, both of observation and observed regularities in celestial events. At best, one might compare much directly predictive work, in some respects, to weather forecasting and, in others, to actuarial prediction.

Although 'prediction' and 'forecast' are often used interchangeably, one should define a more precise usage:

A forecast is a probabilistic statement, on a relatively high confidence level, about the future. A prediction is an apodictic (non-probabilistic) statement, on an absolute confidence level about the future . . . The 'future' referred to in these notions includes situations, events, attitudes, etc.[10]

In general terms, it is rather obvious that the more objectively measurable the boundary conditions of events, and the shorter the time range of predicted or forecasted behaviours, the higher the probability of successful forecasting.

The more complex the events, the greater the number of factors and circumstances that have to be taken into account. The degree to which human interventions and decisions have an effect increases the margin of indeterminacy and reduces the level of predictability.[11]

There are also the phenomena of self-defeating and self-fulfilling predictions. In the former, where some negative outcome is forecast, behaviour may be changed so that this particular outcome is avoided—the prediction fails. In the latter cases, where a positive result, event, or outcome, is predicted, this acts as a reinforcement for those actions that cause the predicted event to come about. In the operation of these factors, feedback from the forecasted future alters the present. This point is one of the central features of what is now called 'normative planning', in which the forecasting and long range planning activity becomes a heuristic or learning function. One starts by positing a future state as it should be, then successively adjusts plans and actions until the preferred future is achieved:

The possibility of acting upon present reality by starting from an imagined or anticipated future situation affords great freedom to the decision-maker while at the same time providing him with better controls with which to guide events. Thus planning becomes in the true sense 'futures-creative' and the very fact of anticipating becomes causative of action. It is at this point that the policy maker-planner is able to free himself from what René Dubos has called the logical future and operate in the light of a 'willed future'.[12]

Success or failure may vary, therefore, in terms of the ways in which these factors operate in different fields of activity. Forecasting methods that work well in one field may be relatively useless in another.

The current approaches to the study of the future might be characterized as follows in summary form:
1. Descriptive—including conjectural, speculative and imaginable modes, etc., as in many classical utopian futures.
2. Exploratory—forecasting based on the methodical and relatively linear extrapolation of past and present events or developments into the future, i.e. the 'logical future', including most technological forecasting, some scenario building and the more deterministic types of socio-economic forecasting.
3. Prescriptive—normatively oriented projections of the future in which explicit value assertions and choices are made about how a specific future may be viewed and achieved.

The level of 'success' in these modes depends on the field and method, as mentioned above. Technological forecasting, where it is restricted to a specific set of physical variables and extrapolates on a known development record in the short range, would have a high rate of successful forecasts. But even in this area there are many cases of unexpected breakthroughs and new inventions which have eluded the forecaster. Economic forecasting, even in the short range, is somewhat less successful. Accurate forecasting of social and cultural events and developments is at a very primitive level.

The problem here is that as one goes from the world of technical events, many of whose constraints are in the domain of physical laws, to the socio-cultural world, change becomes much less deterministic. Many recent attempts to import a more deterministic character to the forecasting of events in the social world, such as the 'limits to growth' type of computer modelling, have assumed a congruence between physical and social processes that may be more imaginary than real.

Indeed one might suggest that prediction or forecasting may not be central to the study of social and cultural futures. In forecasting it is assumed that there is a given set of definable causal relationships between events that may enable one to predict their future state—within varying degrees of probability. Forecasting, therefore, does not usually extend to questions regarding structural premises or implicit 'value' assumptions about the world, which might underlie the basis for forecasting and in which the nature of causality itself may be a prime focus for inquiry. These specific questions are often the major point of departure for the study of social and cultural futures.[13]

One danger to be avoided, therefore, is that of 'accept[ing] uncritically as constants—as socio-cultural universals or "natural laws" for social life—that which may, in fact, represent little more than the outcomes of the socio-historical moment'.[14] This case against the importing or assumption of too much supposedly structural regularity to historical development as a guide to the future is also made by Goldthorpe in a manner that is relevant to our immediate concern with time and the future sense. Referring to ideas such as the 'structure of history' and 'evolutionary movement', he emphasizes:

If these notions are intended to imply that the course of history is shaped by certain invariant regularities or 'principles' which it is possible to comprehend in the form of a scientific theory, then, as Popper has cogently argued, a basic misunderstanding of the nature of science is revealed. The objectives of scientific inquiry may properly be either theoretical or historical—but not both simultaneously . . . In other words, no scientific means exists of arriving at an understanding of scientific events 'as a whole' such that its unfolding to date may be systematically explained and its future course predicted: in science, predictions are conditional statements.[15]

Cognizance of some of these problems is demonstrated in a marked shift in the futures studies field in recent years. There has been a movement away from prediction and forecasting per se towards more normative modes of assessing and projecting alternative future states resultant upon our choices and actions.

One specific example, however, of continuing theoretical weakness in the field is the assumption of time, in the social and cultural sense, as an invariant and universal constant. There has been little awareness of the variability and plasticity of experiential time, and few explorations or projections of the effects of changing concepts of time on social groups and individual lives.

Organizational impacts of futures studies

In the past ten years the formal study of the future has expanded considerably, both in the numbers of persons and organizations at work and in the range, volume and types of activities in which they are engaged.

Reflexively this aspect may have more effect on the future than the success or failure of 'prediction' or 'forecasting'. Just as our physical activities, by the nature of their scale, level of investment, and longer-range impacts, amortize areas of future time and space, our intellectual preoccupation with the future tends to shape and 'colonize' the future.

One index of field growth is the size of international 'future' congresses. The first, in Oslo, 1967, had less than 50 participants; the second, in Kyoto, 1970, had over 250; the third, in Bucharest, 1971, over 300 participants. The World Future Society, a broadly based United States membership society, had over 2,500 people at its 1975 annual meeting.

The increase in the number and frequency of such meetings would be even more dramatic if one included the growth of specialized conferences in long-range planning, technological forecasting, technology assessment, etc., and subsections of the annual conferences in anthropology, sociology, engineering, business, political science, communications and so on, which have 'futures' sections on their agenda.

The perception of world crises in population growth, energy, food and materials supply—and the achievement of the moon landing—fuelled by media coverage of their future dimensions, have turned many shorter-range preoccupations towards longer-term perspectives. This has been reinforced by the emergence of large-scale studies such as 'limits to growth' and by numerous best sellers on the more catastrophic aspects of the future.

Where the earlier predominance of professionals in the field was in the physical sciences, engineering and mathematics, more recent expansion in the past five years has come from the social and behavioural sciences and humanities. This has been accompanied by a significant increase in non-professional participation, which now takes on the proportions of a social movement with many large associational groupings linked to other issue-oriented activities, such as environmental protection, population control and the like.

There is also a growing number of commissions, study groups and long-range 'think-tanks' and 'look-out' operations at international and national governmental levels.[16] Many of these initiatives have developed out of the various United Nations sponsored world conferences on environment, food and population. Although governments have been committed to longer-term planning programmes of various kinds for many years, these have usually been primarily economic and military in focus. In some cases, the recent extension of such planning and longer-term concerns into other spheres is, in part, a direct outgrowth of the futures movement.

These developments in 'the future sense' have strengthened the institutionalization and professional respectability of the field. The recognition of orthodoxy is demonstrated in the appearance of futures studies as a viable subject within educational curricula at various levels, mostly in the developed countries. There is an accompanying drive towards appropriate credentials and status.

Concluding notes

It would seem, therefore, that 'the future of the future' as an object for major societal concern and academic study is likely to be with us for some time. This flowering of interest, coupled with large-scale investment and wider participation in longer-range activities, appears to reflect a major change in temporal perspectives. Whether these tendencies will hold over a long period is, of course, open to conjecture and speculation in itself.

The 'structural imperatives' inherent in many of our present technical and social activities seem to enforce careful projection of their longer-term consequences and implications ever further into the future. Many of our social and environmental problems appear to require the projected commitment of relatively enormous investments of human and physical energies over longer and longer time spans; our energy demands, for example, imply the secure storage of radioactive wastes as a charge on future generations.

When such projection is linked to large-scale social, economic and technological planning and translated into binding political legislation, the overall connotations become more questionable. In the commitment of present resources to ever longer-range goals and purposes we may need to give more careful consideration to amortizing the future in terms of what may be wholly temporal constraints. To what degree can we justify the sacrifice of present time to the more speculative exigencies of the future? The further adumbration of the future sense might well include the protection of the present from the apparent future demands—or, reciprocally, the conservation of the future from untoward erosion by the present.

NOTES

1. J. T. Fraser (ed.), *The Voices of Time*, p. 598, New York, George Braziller, 1966.
2. Mircea Eliade, *The Sacred and the Profane*, p. 95, New York, Harper and Row, 1961.
3. John Cohen, 'Subjective Time', Fraser, op. cit., p. 262.
4. Mircea Eliade, *The Myth of the Eternal Return*, p. 104, New York, Bollinjen Foundation, 1954.
5. Max Weber, *Essays in Sociology*, trans. by H. H. Gerth and C. W. Mills (eds.), p. 275, New York, Oxford University Press, 1958.
6. J. B. Bury, *The Idea of Progress*, p. 194, London, Macmillan, 1928.
7. Lewis Mumford, *Technics and Civilisation*, p. 14, New York, Harcourt Brace and World, 1963.
8. Stephen Toulmin, 'The Problem of the Time Barrier', *The Listener*, 21 January 1965, p. 97.
9. A similar shift is now being experienced with the measurement of time. For example, though the unaided human sense cannot distinguish less than 1/100 of second, computers now operate on nanosecond intervals, with one billion nanoseconds to the ordinary second. Many of the crucial timing operations in society have now slipped below the sense threshold in this way and may only be apprehended via instrumented means—much in the same way that we use microscopes, meters and other instruments to monitor and control below and above our other sensing thresholds. Radioactive-decay dating techniques, at the other end of the scale, are the long-range telescopes of our time sense.
10. Eric Jantsch, *Technological Forecasting in Perspective*, p. 15, Paris, OECD, 1967.
11. For a close discussion of this point, see Olaf Helmer and Nicholas Rescher, *On the Epistemology of the Inexact Sciences*, New York, Rand Corporation, Report No. 353/1960.
12. Hasan Ozbekhan, 'The Triumph of Technology: "Can" Implies "Ought"', Systems Development Corporation, SP-2830, June, 1967.
13. This area of divergence is discussed in more detail in John McHale, 'Society', *Forecasting and Futures Research*, Vol. 12, No. 5, July/August, 1975.
14. William Simon, 'Reflections on the Relationship between the Individual and Society', *Human Futures*, p. 145, London, Futures, IPC Science and Technology Press Ltd, 1974.
15. John H. Goldthorpe, 'Theories of Industrial Society: Reflections on the recrudescence of historicism and the future of futurology', International Sociological Association, 7th World Congress, Varha (Bulgaria), September 1970.
16. More detailed data on these organizational aspects of the field may be found in John McHale and Magda Cordell McHale, *Futures Studies: An International Survey*, New York, United Nations Institute for Training and Research, 1975.

EXPRESSIONS OF TIME
IN INFORMATION SCIENCE
AND THEIR IMPLICATIONS

A. Neelameghan

Introduction

In a consideration of the expressions of the time parameter in information-science studies, three factors appear to be interesting: (a) the qualitative and quantitative expressions of time in the information field, based on analogues and models largely borrowed from other disciplines; (b) time as a unifying and correlating concept among a variety of disciplines, including information science; and (c) changing patterns and developments in the information field, and their implications to, and impact on, individual and group behaviour.

Time-binding and communication

THE SOCIAL FUNCTION OF COMMUNICATION

The social function of communication is the ensuring of continuity in society through access to the experiences and ideas of the past, expressed in symbols for transmission across space and through time. This is the 'time-binding' function of social communication. In the 1920s Alfred Korzybski[1] proposed a theory of organic and human behaviour, based on binding ability, which provides an insight into the links and differentiation between plant, animal and man. During recent decades, studies in cybernetics, thermodynamics, ethology, neurobiology, etc. have provided supporting evidence and data for the theory.

Man's time-binding ability, arising from his usage of language, number, gesture, picture and other symbolic forms, enables him to transcend the limitations of inherited characteristics and the seemingly insurmountable barriers of 'time' and 'space'. The insight, the wisdom, the culture and the arts, the ideals of the political, economic and social structures of the world, and the human values and experiences of past generations, can be preserved, transmitted and used and evaluated by man both in the present and in the future thanks to the time-binding, culture-binding mechanism of symbolic communication. Implied in this is the 'culture concept' expressed by anthropologists, linguists and sociologists. An animal lives directly in the 'present', without any apparent notion of the animals of its kind that have lived before it, and of those that may come after it. Man, on the other hand, because of his capacity for symbolic communication, has a sense of events in sequence—i.e. a sense of time. From this arises his motivation to transcend time, as evidenced by the great depth and breadth of his cultural borrowing from the past and lending to the future. Man's superiority in the competition for self-perpetuation arises from his ability to use signs to indicate entities and to represent them. McGarry points out: 'The use of these substitute signs is the essence of symbolism and our power of reasoning; they can take the place of past experiences and feed the imagination by combining memories of things that might be in future experience; such an experience takes man from the here and now.'[2] Further, the ability to transfer information received by one sense mode to that of another—that is, cross-modal association—appears to be unique to man. It endows him with special creative ability to plan ahead and influence the future.

SURVIVAL IN TIME AND CULTURAL CO-OPERATION

Survival for man implies three facets: survival in energy, survival in space, and survival in time. It involves both physical inheritance and competition as well as cultural inheritance and competition—that is, competition in, with and for symbolic forms used in communication. Intra-species competition reduces the chances of survival of the species. Therefore, co-operation within species is essential to the physical survival and time-binding culture-transmission activity of man. Man is a link in the vast transmission line of the culture through time.

Knowledge and culture transmission and the perpetuation in time of society are secured not only through oral communication, but also through various social structures and forms: ' . . . rites and rituals, sacraments, ceremonies, and institutions. Language is itself one such institution and the continuity between speech, writing, print, books and libraries defines language's time-binding institutional function'.[3]

Societal homoeostasis

INFORMATION AND ENTROPY

Human society can be looked upon as a dynamic open system consisting of individuals, groups, organizations and institutions of various kinds in continual interaction and interchange of energy and messages (information). Here we shall primarily consider semantic interchange. Open systems have certain common characteristics, such as the following:[4]

Input. Import of energy from the environment. Man, as a biological and social entity, seeks and inputs energy and the information necessary to satisfy his biological, emotional, intellectual, spiritual, economic and cultural needs.

Throughput. The input is processed by performing work and is converted into the output. Man himself is an information-processing system. He has also created various information-processing systems to help him.

Output. Inference, data, ideas, description, information, knowledge, documents, etc.

Cyclic character. Input, processing, output, feedback, control, adjustment and adaptation—each phase supports the next in a cyclical way. From this succession of events in time arises the interrelationship of the pasts of the system and of their functions.

Negative entropy. According to the second law of thermodynamics, the universe is running down. The energy loss or increase in entropy corresponds to the positive direction of time. Thus time's arrow, or probability vector, affects the physical, biological and cummunicative patterns in the universe.

Some of Korzybski's time-binding agents, man in particular, constitute these local enclaves trying to maintain order. The commands through which man exercises control over his environment are a kind of information imparted to it. Increase in the supply of information increases negative entropy, helps to reduce uncertainty, imposes order, and prevents the system from tending towards chaos and disorganization.

Information input, negative feedback, and coding

Negative feedback is the information on what, if anything, is wrong with the system, on why it is necessary to make adjustments and take corrective action. Signals received may be expressed as codes. Man has evolved elaborate coding systems to facilitate communication.

Steady state and dynamic homoeostasis

The change in the attributes of the parts of the system over a time period t_1 to t_2 may not be drastic (steady state). On the other hand, information received may cause substantial impact, even cognitive dissonance, on the individual.

Just as the human body is kept in dynamic equilibrium—that is, homoeostasis—through cell co-operation and communication, so also human societies depend upon cultural communication for their homoeostasis. Therefore, co-operative behaviour patterns and the resulting meaningful effective human relations have a high survival value. And these behavioural patterns are based on the act of symbolic communication.

Differentiation and equifinality

Division of labour and specialization of functions and roles of the different parts of the society are necessitated by the growth in time of the size of the society and the complexity of its demands.

The possibility exists of attaining the set goals of the system, starting from a variety of initial conditions, by adopting suitable alternative means. The different phases of the system's working are time-dependent, and processes/events, such as entropy and steady state, are referred/defined specifically in terms of time and information communication.

SOCIETAL COHESION AND COMMUNICATION

Human civilization and culture, organizations and institutions are held together over time in a network of mutual dependence and responsibility and the sustaining transactions are carried on by myriads of messages communicated over this network. These messages form the basis as well as the regulating, organizing and time-binding influence in social evolution and progress. A failure in communication could lead to misunderstanding, misinterpretation and confusion of the import of the messages.[5] In turn, these may cause alienation, aggression and detraction from human progress. Here communication connotes a deep, dynamic and complex series of processes operating together—perception, memory, information analysis and processing, symbolic transformation, delivery and feedback. The blending of the resources of cybernetics, information theory, systems approach and semantics with those of science and technology has extended human foresight. Man, thus, appears to be moving toward influencing the future and directing his own evolution.

Time in the communication process and knowledge growth and diffusion

GENERAL SYSTEMS APPROACH

As an approach to the organization of general systems theory, Kenneth Boulding suggested the need 'to look over the empirical universe and to pick out certain general phenomena which are found in many different disciplines, and to seek to build up general theoretical models relevant to these phenomena'.[6] Examples of such common phenomena are population dynamics, growth in time, relative growth, entropy, competition, information diffusion and so on.

GROWTH IN TIME: EXPONENTIAL AND LOGISTIC PATTERNS

Patterns of growth in time showing isomorphism of law in different fields are the exponential and the logistic. Examples are the growth of science and technology, the increase in knowledge of the number of animal species, yeast cells in a controlled environment, the increase in publications on drosophila, in manufacturing companies, and so on. Isenson[7] has shown that the rate of growth of knowledge over time can be expressed as an equation embodying various limiting factors such as the upper bounds imposed by natural law, political and social constraints, the contemporary level of scientific and technical understanding, and the ability of the scientific and technical communities to communicate their new findings.

As several studies have shown, the amount of recorded knowledge (literature) in many scientific fields doubles, on the average, about every ten years; but exponential growth must have ceilings because of various constraining factors, otherwise the growth would reach absurd conditions in the future.

The logistic growth curve may not be a normal S-shaped ogive. Historical time series analyses present variants of the curve as seen for example, in the growth of science and technology. Hunting fluctuations and the escalating curve are two examples. The constraining factors causing the variant shapes, and the implications to the sociology of science, have been the subject of study by De Solla Price.[8]

TECHNOLOGY GROWTH: THE BOLTZMAN EQUATION MODEL

Hartman found an analogy between nuclear chain reaction and technology growth. He has suggested that a model for the technology forecast can be derived directly from the Boltzman equation.[9] Scientists and engineers are pictured as being immersed in a sea of information, the information being a necessary input for the generation of new information (technology). His thermodynamic model simulates authors by stationary atoms and information carriers by electrons accelerated by an electrical field. A bit of information may collide with a researcher and trigger a new idea or it may not. As it is not possible to predict what piece of information will collide with which researcher and whether a particular collision will be 'elastic' or 'inelastic', the process is only statistically definable—providing an acceptable parallel for forecasting technology growth.

INFORMATION DIFFUSION: THE HEAT CONDUCTION MODEL

Avramescu describes[10] a mathematical model for scientific information transfer on the analogy of heat-diffusion laws. The information-flow equation is similar to Fourier's. The information-diffusion process is a multiple chain, one scientific paper stimulating another. The elementary space elements are papers linked together by references and citations in a natural time sequence.

IDEA ACCEPTANCE: TIME-DEPENDENT MODELS

Related to the concept of diffusion of ideas is the concept of acceptance of ideas by a target group. Morphet[11] has presented a probabilistic model of time-dependent factors governing the acceptance of innovative ideas by industry. Within this framework is analysed the timeliness factor that partially governs the receptivity to innovative ideas of industrial organizations.

The model is based on the assumptions that: (a) even the best innovations do not diffuse by themselves; (b) the commitment of an organization to existing practice will favour innovation only when existing practices become obsolete. The occurrence of such obsoleteness is identified with the concept of the New Idea Point (NIP), the point in time at which the organization must have assembled a portfolio of innovative ideas in order to maintain continued existence. The reaction of firms to new ideas is associated with the organizational ambivalence—a tension between action and reaction, between need and conservatism—which must be periodically resolved in order to meet the imperatives of the NIP.

INNOVATION DIFFUSION MODELS
AND THEIR SOCIOLOGICAL IMPLICATIONS

Everett Rogers and others have attempted a generalization of the mode of diffusion of research results in the 'market' over a period of time. An innovation is defined as an idea that is seen as new by an individual. Thus, innovations include a weed spray (2.4-D) among farmers, a drug among physicians, the intra-uterine contraceptive device among peasant women in India, an education technique among teachers, and so on. The crucial elements in the diffusion of innovations are deemed to be the new idea, the communication channels, the particular social system, and time. The element of time distinguishes the diffusion studies from other types of communication studies. Time is involved in the following ways: (a) the innovation decision period through which an individual moves from first knowledge of the innovation to persuasion of its usefulness, to its adoption, and to continued use; (b) the rate of adoption of the innovation in a social system, measured as the number of adopters per time period; and (c) the innovativeness or the degree to which an individual is relatively earlier than other members of his social system to adopt new ideas. Factors helpful in accelerating the diffusion and adoption of new ideas/innovations, such as altitudinal type variables, social relationships and strategies adopted by change agents, have also been studied.[12]

THE ACCELERATION OF INNOVATION
AND KNOWLEDGE OBSOLESCENCE

The Second World War and the events that followed it provided a great impetus for scientific research, innovation and technological change, which in turn contributed in a large measure to socio-economic change. Some of these innovations, particularly in communication and information technologies, made possible greater interaction among researchers and innovators, which in turn further accelerated the scope and pace of scientific and technological inquiry and innovation. It is estimated that half the total of scientific research in the United States has been done since 1965. Similarly, in the United Kingdom since the Second World War more than 50 per cent of all new products have been based on recent scientific and technological discoveries and innovations.

In addition to the expansion in the scope of scientific inquiry, there has been an appreciable acceleration in the rate at which new knowledge is put to use. This has resulted in an acceleration in the rate of knowledge obsolescence. The United States Department of Commerce has estimated that prior to 1914 there was an average wait of thirty-three years between an invention and its application. By the Second World War, the time lag had dropped to ten years and now it is even less. According to some findings of a study by Killingworth and Lynn (a) the time lapse between innovation and its utilization in industry is much shorter for consumer products than for industrial products; (b) the time lapse is shorter for invention developed using government funds than for those using private funds; and (c) the technology diffusion and transfer time is reduced if the innovator himself undertakes the development.

What are the implications? A major change that might have required five years to make a decade ago must now be completed in a shorter time period, if an organization, particularly an industrial or commercial organization, is to remain competitive. On the one hand management reaction time is constantly shrinking, while on the other hand each decision may involve more risk (because of a large number of variables, internal and external, to be taken into account), and is valid only for a shorter time span. Therefore, as reaction time diminishes, opportunity for gainful action may be lost, because preoccupied managers and policymakers may fail to reach out and grasp them.

Time and information systems

ZIPF AND BRADFORD DISTRIBUTIONS

In bibliometrics since 1960, two classes of empirical laws have been used fairly frequently. These are the Zipf and Bradford laws,[13] both of which are hyperbolic distributions. One or the other of the laws is used, depending upon whether one's interest is in vocabulary, periodical literature, physical access time, the rate of diminishing returns in bibliographical search, or the cumulative yield from a given input. The time factor involved here is in the explanation of the observed relation in many fields on the basis of the general principle of least effort.

DOCUMENT LIFE PARAMETERS

It has been observed that if all the citations in a single issue of a periodical, or a volume of it for a particular year, are sorted according to the date of the cited documents, then the number falls off rapidly as we move back into time. Since 1960 several studies of this type in different subject fields have been done.[14] From these studies several time-related parameters of a periodical's life have been isolated, such as 'half-life'—the time period, actual or expected, during which half the total use of an individual item constituting a literature has been, or is expected to be, made.

The values of such parameters can be used for guiding library management decisions: e.g. maintenance of library collection (periodicals, books, etc.), withdrawal of documents from active circulation, planning of acquisition programmes within a hierarchy of library systems, and identifying significant research in a field.

WAITING TIME

In the management of any system including an information system, ensuring receipt of the item sought by the clientele with minimum delay and cost is an important consideration. The 'waiting time'—the subjective or psychological time—is of special concern here. For users of an information system, this may arise in one or more of the following stages: (a) physical/communication access to the system; (b) identification of and access to the appropriate data base; (c) use of the tools and techniques to search in the data base to select and retrieve the relevant information/surrogate; (d) document information delivery; and (e) document/information usability.

A variety of improvements are being progressively incorporated into information systems, manual and mechanized, to secure quick access to the relevant information and documents.[15]

Research communication: sociological and behavioural aspects

OVEREMPHASIS ON SPEED?

Timely availability of relevant and reliable information and data is important to the researcher and the production engineer, as well as to the manager and planner, as an aid in solving problems, making gainful decisions, minimizing the chances of unnecessary duplication of effort, and triggering new ideas. With the growing demand for information and data, there has been a phenomenal growth in primary sources such as books, periodicals, reports, conference proceedings, and so on. In finding ways and means of moving the information from the point of generation to points where it will be of use, the need for speed of communication has been emphasized. For instance, reducing the time lag between submission of a paper to a periodical and its publication has been an important consideration. Growth in the number of primary periodicals, publication of short communications, minimizing the editorial work, and the adoption of fast printing techniques are some of the results of such considerations. There have been salutary effects as well as the undesirable effects of the ever-increasing quantity of information that swamps the user at an apparently ever-increasing speed.

Ziman points out that in the pure sciences the supposed wastage of resources through duplication of effort is somewhat exaggerated. Simultaneous discovery arising from concurrent work on the same problem by different researchers is not a bad thing in itself. It provides for a good means of confirmation of the findings or discoveries; the routes taken to the discoveries by the different researchers may be different. Experiments may be repeated to improve techniques or increase accuracy of data. Also, the 'speed of research' varies at different stages, from discipline to discipline, and the 'speed of supply of information' should correspond to this variation in demand.

In spite of this, the rate of proliferation of primary publications and of information exchange media has gone on unabated. The 'publish or perish' syndrome is bound up with many questions in the sociology of science (i.e. the achievements of a scientist, especially the newcomer to a field, being estimated on the basis of the number of papers published by him, the economics of research and the rat-race for grants, the behavioural aspects of researchers to 'scoop the field' as if in competitive business to ensure priority of 'discovery', which in turn could bring more projects and grants), the economics of the information industry, and so on.

The sociological and human aspect of science also throws some light on the motivations of the scientist for quick announcement and speedy communication of his work. The motivation is to establish and maintain

intellectual property. Duplication of discoveries is a widespread phenomenon.

The issues discussed above relate to communications in 'pure'
sciences. The situation is somewhat different in the applied and development research areas. Here one seeks the best available solution to a
particular problem, and the rewards of success or cost of failure may be
measured in thousands of dollars. Timely information and data helpful in
making gainful decisions are vital in competitive industrial research. Here
too, however, when users are research managers, production managers,
etc., the information system should not only be capable of picking out
quickly the most relevant data and information, but also present them in
a manner—i.e. digest, technical note, trend analysis with charts and
graphs—that will immediately show the relevance of the information to
the activities of the organization or the individual concerned. Such
repackaging of the information and data would facilitate the reaction of
the individual to the perceived situation in the context of a rapidly
decreasing management reaction time to environmental changes.

Ziman points out that 'hindrances to the immediate spread of new
knowledge (in applied and development research) are not then so much in
the machinery of publications, but rather barriers of secrecy deliberately
raised around industrial and military research. The communication
system of technology is quite different from that of pure science, having
different ends, different norms, and altogether different standards of
morality'.[16]

BEHAVIOURAL ASPECTS

Kenneth Arrow and others have been examining how the amount of
information processed by businessmen, consumers, stockholders and
other 'economic actors' influences their behaviour on the stage of the
economic system. Using such considerations and looking upon information as a key resource, the 'information revolution' is influencing the
pattern of thought in several fields—economics, political science, government, sociology, and so on. Economists are attempting to formulate a new
economics of information, and over the past decade it has ranged over a
wide spectrum of activities in the information field—from the study of
specific activities of libraries to a reconsideration of the role of
information in the economic system,[17] and the welfare economics of
symbol manipulation.[18]

The emerging information environment

The acceleration of the pace of life (i.e. speed-up of production through
automation) means that every minute of 'down time' costs more in lost

output than ever before. Information must flow faster than ever before. At the same time rapid change, by increasing the number of unexpected problems, increases the amount of information needed. This combined demand for more information at faster speeds is cracking the rigid vertical hierarchies of typical bureaucracies.

New knowledge either extends or outmodes the old. In either case, it compels those for whom it is relevant to re-learn today what they thought they knew yesterday. Robert Hilliard, a broadcasting specialist for the United States Federal Communications Commission, comments:

At the rate at which knowledge is growing, by the time the child born today graduates from college the amount of knowledge in the world will be four times as great. By the time the same child is fifty years old, it will be thirty-two times as great, and 97 per cent of everything known in the world will have been learned since the time he was born.

Max Weber points out that the remarkable increase in the speed by which public announcements as well as economic and political facts are transmitted exerts a steady and sharp pressure in the direction of speeding up the tempo of administrative action.

In the information-rich and technologically fast-developing world of tomorrow the pace and direction of social change is likely to be set and guided by those individuals, institutions and nations that have the capabilities for optimal use of information in making gainful decisions in all areas of human endeavour.[19]

Barriers to communication: temporal dimensions and social consequences

INFORMATION AND GOAL ACHIEVEMENT

In designing information systems a basic premise is that economic, social and political systems will perform better and more efficiently if a mechanism is provided to ensure that decision-making points have timely access to relevant, reliable and adequate information. Therefore obstacles to convenient access to information and absorption of ideas—for example, barriers to communication—can lead to delay in the achievement of social goals.

INFLUENCING FACTORS

Communication of ideas, the basic behavioural act of man, has helped to accelerate change in almost every sphere of activity designed to satisfy his wants of one kind or the other. In fact, the technological achievements since the Second World War have been so dramatic that people have been able to share the hope that they need no more go hungry, ill-clad, or unsheltered owing to ignorance—that is, inaccessibility, or delay in the access, to knowledge and the means of its utilization. Yet it is apparent that vast numbers of people are living close to misery, fear, tension and conflict. Programmes for the sharing of the affluence and misery between the haves and have-nots have, in many cases, failed to produce the desired result. Here again, inadequate information, delayed information, misunderstanding of information and misuse of information have been among the principal causes of the failures.

Attributes of the social system

The attributes of the social system as a whole that may influence the effectiveness and speed of transfer of ideas, and therefore of culture among people, include (a) whether the situation involves one-to-one or one-to-many communication; (b) whether the communication takes place within national boundaries or across national boundaries; (c) whether the transfer of culture is attempted among equals or non-equals; and (d) the total scientific, social, political and economic environment, present or past.

Attributes of participants

These may be (a) the objectives and motivations of the participants involved; (b) the differences in religious, educational, economic, social and cultural backgrounds among the participants; (c) the intimacy of contact; (d) prior experiences; and (e) differences in information richness.

Attributes of medium and mode of communication

These are (a) whether an intermediary is involved and (b) whether it is an individual, institutional or machine system.

TAXONOMY OF OBJECTIVES AND COMMUNICATION
ACROSS CULTURES: THE TIME PARAMETER

In a scientific inquiry into an object the usual question to which an answer is sought is 'What is it?' Questions asked here include the following time-questions: When did or does it occur? How long did or does it last? Which time-slice is of particular interest?

The answers to these questions are expected to increase the inquirer's understanding of the object under study. These parameters are also used in developing a taxonomy or classification of objects and to organize information about the objects. Such taxonomy and organization of information are deemed to assist effective communication with those who seek information about the objects. In most of the modern information-classification systems 'time' is used as a parameter for categorization, together with all the relevant and related qualitative and quantitative expressions of time, e.g. past, present, future, early, late, 2 o'clock, 1975, Gregorian calendar, and so on. But there are cultures whose concepts of time may be in complete variance with this sort of conception. (See the papers in this volume by Kiray and Abou-Zeid.) The question is often raised as to how compatibility can be achieved between two different taxonomies of time such that cross-cultural communication of information can be made effective.

The 'silent language of time' used in different cultures in different ways to communicate a person's or a group's reactions to environmental situations has been eloquently and analytically dealt with by E. T. Hall.[20]

Particular taxonomies arise according to needs. For example, in some of the South Sea island communities not all the seven colours with which we are familiar are recognized, for they do not need all of them for the conduct of their daily activities. It is also known that as some of the community members became 'modernized' and shifted to another cultural context, their communication repertoire absorbed additional colour concepts and vocabulary. Therefore, attempts to introduce a new idea or an innovation without a proper understanding of these socio-cultural and psychological characteristics of the target group, and also the attributes of the communication system, can delay socio-economic change in the desired direction—and the cure may even prove to be worse than the disease.

The future: some research issues

In an increasing measure it is being realized that world peace and prosperity require the creation of a new international economic and social order—something different from the one that obtains now, marked by dichotomies, distinctions and discriminations between the haves and the have-nots, rich and poor, advantaged and disadvantaged, privileged and underprivileged, developed and underdeveloped, and so on. The creation of this new economic and social order depends on the co-operation and collaboration of nations, institutions of all kinds, and different social groups, so as to bring about the appropriate changes in attitudes of people as well as in social structures, which hopefully would reduce the gaps between cultures, nations and social groups. Knowledge is the principal instrument for social and economic change, and the sharing of knowledge could be an effective means of reducing and bridging the gaps—spatial, temporal, cultural—in and between societies. The capability for knowledge transfer and information handling is both the potential for change as well as the indicator of achievement in material wealth. As the second report of the Club of Rome puts it:

All contemporary experience points to the reality of an emerging world system in the widest sense which demands that all actions on major issues anywhere in the world be taken in a global context and with full consideration of multidisciplinary aspects. Moreover, due to the extended dynamics of the world system and the magnitude of current and future change, such actions have to be anticipatory so that adequate remedies can become operational before the crises evolve to their full scope and force. If actions are to be anticipatory and effective, they must be based on a supply of information which is as complete and accurate as it can be made to be.

Will the emerging trends in the information field—e.g. the new economics of information based on the concept of information as the major non-depleting resource of society, the profound mutual impact of society and information, the shifts in the loci of decision-making and the exercise of power—lead to the new economic and social order the United Nations hopes for? Or will the changes merely create new cultures and a shift in the balance of power, and precipitate new disparities in society? It is prophesied that everyone will benefit from the coming change propelled by increased access to information; it is a non-zero-sum game. Will the reduction of gaps in society lead to a uniformization and averaging out? Or will there be adequate scope for the development of the individual, his personality and creative ability? Will the creativity be achieved through better use of the 'grey cells' of the human brain—use of a larger proportion and for a longer period of time of the individual's life-span? Will these be alternatives or complementary technologies useful in different contexts? And will increased creativity result in less or more use of information per individual on the average?

Whether the kind of society that will emerge say in the next twenty-five years will be the most desirable kind remains to be seen. In any case, in the coming decades a good part of the world's efforts will be directed towards devising means and methods of increasing access to and use of information at all levels. Will the developing countries benefit from these efforts? Should they use the same technologies that are now being developed in the technologically advanced countries? Or should they look for complementary and alternative methods to achieve the goals they set for themselves?

Given the fact that whereas the totality of production, storage, organization, distribution and utilization of information is a long-range and expensive business, but that information is the basis of gainful decisions, planning, and control at all levels of human activity, the correct answers to the questions raised in this paper could provide the elements for formulating an objective and integrated information policy.

NOTES

1. Alfred Korzybski, 'Time-binding: the General Theory', presented before the International Mathematical Congress, 1924 Toronto.
2. Kevin J. McGarry, *Communication Knowledge and the Librarian*, p. 46, London, Clive-Bingley, 1975.
3. George M. Evica, 'Time's Arrow and the World's Order: a Re-evaluation of Korzybski's Binding Theory', *ETC: A Review of General Semantics*, Vol. 20, No. 4, May 1963, p. 421–37.
4. Ludwing von Bertalanffy, 'General Systems Theory—a Critical Review', in Walter Buckley (ed.), *Modern Systems Research for the Behavioral Scientist*, p. 11–30, Chicago, Aldrine Publishing Co., 1968.
5. A. Neelameghan, 'Social Change, Communication of Ideas, and Library Service with Special Reference to Developing Countries', *Library Science with a Slant to Documentation* (Bangalore), Vol. 10, No. 1, March 1973, p. 1–29.
6. Kenneth E. Boulding, 'General Systems Theory—the Skeleton of Science', in Buckley, op. cit., p. 3–10.
7. R. S. Isenson, 'Technological Forecasting in Perspective', *Management Science* (Providence, R. I.), October 1966, p. 370–83.
8. Derek J. De Solla Price, *Little Science, Big Science*, p. 23–4, New York, Columbia University Press, 1963. (The George B. Pegram Lectures.)
9. L. M. Hartman, *The Prospect of Forecasting Technology*, p. 237, New York, Macmillan Co., 1966.
10. A. Avramescu, 'Modelling Scientific Information Transfer', *International Forum on Informatum and Documentation*, Vol. 1, No. 1, 1975, p. 13–19.
11. C. S. Morphet, 'A Probabilistic Model of Time-dependent Factors Governing the Acceptance of Innovative Ideas by Industry', *R & D Management*, Vol. 4, No. 3, 1974, p. 165–75.
12. Everett M. Rogers and J. David Stanfield, 'Adoption and Diffusion of New Products: merging Generalization and Hypothesis, in Frank M. Bass, Charles W. King and Edgar A. Pessemier, *Applications of the Sciences in Marketing Management*, p. 227–50, New York, John Wiley & Sons, Inc., 1968.
13. George K. Zipf, *Human Behaviour and the Principle of Least Effort*, New York, Hafner, 1949.
14. R. E. Burton and R. W. Kebier, 'The "Half-life" of Some Scientific and Technical Literatures', *American Documentation*, Vol. 11, No. 1, January 1960, p. 18–22.
15. A. Neelameghan and S. Seetharama, 'Information Transfer: the Next Twenty-five Years: Librarianship and Information Science in 2001', seminar arranged by the British High Commission, British Council Library, on the occasion of its Silver Jubilee, Madras, 7–10 October 1975. p. 9–37.
16. J. M. Ziman, 'Information, Communication, Knowledge', *Nature*, Vol. 224, 25 October 1969.
17. J. Hirschleifer, 'Private Social Value of Information and the Reward to Inventive Activity', *American Economic Review*, Vol. 61, September 1971, p. 561–74.

18. J. Marschak, 'Economics of Enquiring, Communicating and Deciding', *American Economic Review*, Vol. 58, May 1968, p. 1–16.
19. Conference Board (New York), *Information Technology. Some Initial Implications for Decision-makers*, New York, The Conference Board, 1972.
20. Edward T. Hall, *The Silent Language*, New Delhi, Affiliated East-West Press Pvt. Ltd, 1959.

THE CONCEPT OF TIME IN PEASANT SOCIETY

Ahmed M. Abou-Zeid

Peasant behaviour

A study of the concept of time and systems of time-reckoning in traditional peasant societies is not an easy task. For one thing, time is seldom regarded by villagers in these societies as a mere tool that should be exploited in achieving certain well-defined objectives, or a mere means of measuring activity and productivity. It is also a means of consolidating existing social relationships and establishing new ones. Therefore any serious attempt to study the attitude of traditional peasants towards time should try to analyse the whole network of social relationships and understand the system of values prevailing in these societies. In fact, the relatively few sociologists and anthropologists who have studied the subject have been keen on emphasizing that peasants have, in general, a different outlook, and a different way of evaluating time and its role in organizing their activities and their life, from the outlook of urban people, especially in Western culture. They are sometimes described as people who seldom consider time as something actual, which passes, can be wasted or saved, and so on. Evans-Pritchard once asked whether the pastoralist Nuer 'ever experienced the same feeling of fight against time or of having to co-ordinate activities with an abstract passage of time, because their points of reference are mainly the activities themselves, which are generally of a leisurely character'. Similar ideas can be found in a number of writings about peasants in traditional societies. Peasant behaviour is often described as characterized mainly by indifference to time and not regulated according to a timetable. That is why peasants act and work without haste. 'Haste is seen as a lack of decorum combined with diabolical ambition.' They are masters of passing time, 'or better, of taking one's time'. Bourdieu thinks that the worst discourtesy in the eyes of peasants is 'to come to the point and express oneself in as few words as

119

possible. Sometimes one leaves with nothing settled'. The notion of an exact appointment is not known. People may agree to meet, not at a certain precise hour, but just 'in the morning' or 'after the last prayer'. In Algeria women are often seen waiting at the hospital or before the doctor's office two or three hours before it is due to open.

There is plenty of evidence from peasant societies supporting these statements. Peasants in Egypt, for example, may spend the whole night at the railway station to catch the next morning's train, for, as they explain, one should wait for the train because the train does not wait for anyone. Villagers go to the village café mainly to meet each other. They chat freely for long hours about their daily life and their private problems. It would be discourteous of the waiter to come hurriedly to get their orders. A 'waiter' should 'wait' till he is called by the clients. Actually it is part of his duty to welcome the clients as if they were personal friends and guests, and to exchange friendly talk, including some jokes, with them. Even in economic transactions and exchanges, the same attitude prevails. One goes to the village store not only to buy whatever goods one may need, but also, and perhaps primarily, to 'pay a visit' and have a chat with the storekeeper and other customers. Only at the end of the 'visit' one may, or may not, buy the goods one wanted. Much haggling and bargaining takes place between merchant and client before they agree on the price of even the so-called fixed-price goods. Haggling is regarded as an art. It is a means of communication, and its establishes friendly relationships if conducted in the right way. A client who always knows what he wants, and who always pays the price the merchant asks, and then leaves hastily, is not regarded as the ideal type of customer.

This leisurely attitude towards time may be the outcome of the very nature of agricultural life and activities, which are organized according to seasonal variations, rather than by hours or any other shorter time unit. Thus, time units tend to be less well-defined than they are in urban communities, and more particularly in Western societies. Keeping to a strict and precise timetable is of little significance.

Problems of diversity

Another source of difficulty and confusion is the diversity of systems of time-reckoning used in traditional peasant societies. Most if not all these societies have two calendars at least: a traditional one (sometimes more than one) and the Western calendar, which has been adopted at a more recent date. Each of these calendars has its own functions and is used in its own particular context. Traditional calendars are generally used for regulating the major traditional social activities, whether economic (such as agriculture) or religious. The Western calendar is used for regulating the more 'modern' aspects of daily life, especially those that are connected with the government and semi-governmental activities. Many of these societies follow a lunar calendar in organizing their agricultural and religious activities. Prevalent myths may attribute to the moon phases certain characteristics that have an effect on cultivation. Many observers indeed, believe, the moon was man's first means of keeping track of time; although most historians of astronomy do not agree with this now, on the grounds that the stars are more widely used for time-reckoning than the moon, even in primitive societies. In any case, very little effort is made to co-ordinate these various calendars, or to bring them into one homogeneous system.

Examples

In Egypt, a country with a long history of elaborating time-reckoning, three different calendars are used by the peasants: the Coptic calendar, the Islamic calendar and the Western calendar. The Coptic calendar is a solar calendar, but it does not conform with the Western calendar (which is also solar). The first month of the Coptic calendar, i.e. the month of Toot, begins on the tenth or the eleventh of September. It is most probably based on the Ancient Egyptian system of time-reckoning, and it is still followed in organizing agricultural activities. City dwellers are seldom acquainted with it, or with the names of its months, which are generally related to the significant climatic changes that affect cultivation. The relations between the months, the climatic conditions, the agricultural activities that take place during each month, and the state of cultivation and the crops are spelled out in a number of rhyming phrases to help the uneducated peasant to memorize them. Thus Hatoor, the third month in the calendar, is described as *Hatoor aboul dahab el-mantoor*, or Hatoor the father of the spread-out gold, referring to the fact that wheat is usually sown during that month. Such rhymed phrases and proverbs are sometimes anticipations of the future, since events of agricultural life repeat with a very high degree of regularity. It is the same with the Algerian peasants, whose anticipation 'consists of reading the signs to which tradition furnishes the key'. One of their proverbs says 'If it thunders in January, take up the flute and tambourine; if it thunders in February don't use up your fodder'.

The second traditional calendar in Egypt, the Islamic, is used to organize religious life and ritual ceremonies and festivals. This purely lunar calendar is regarded by many writers as the strangest calendar in wide use today, for its year consists of six 29-day months and six 30-day months, which makes a year of 354 days. The seasons and months have no connection with each other, and the religious activities shift from one month to another in accordance with the Western calendar. The Western calendar itself is used to schedule civil events. Peasants seldom rely on it in their ordinary life, although there is a clear shift now towards using it, even in agricultural activities, by the young peasants who have gone to public schools. This represents an important change in the system of traditional calendars and time-reckoning.

Chinese peasants, to give another example, also have two traditional calendars (one lunar, one solar) in addition to the Western calendar. The traditional lunar system is largely used in 'such institutions as remembering sentimental events and making practical engagements'. Thus it regulates the occasions of visiting temples, the holding of ancestral worship ceremonies, the celebration of vegetation rituals and festivities and many other traditional social activities, although some of these festivals are regulated according to the traditional solar system. The traditional solar system is not used as a system of dates; it is concerned with climatic changes and allows each locality to adjust its calendar of work to local conditions: i.e. it is used mainly in relation to productive work. On the other hand, the Western system is used chiefly in connection with newly introduced institutions such as schools, co-operative factories, administrative offices and so on. These have to adjust their work and activities with those of the outside world where the Western system is used.

This complexity of calendars, found in many other traditional peasant cultures, is likely to cause much confusion and perplexity, more especially for foreigners. Thus the Chinese villagers remember and predict their sequences of work in terms of the traditional solar system. But this system cannot stand alone, because it is very difficult to understand without a well-defined system of dating, and since it is the lunar calendar that provides the system of dating, the villagers find themselves obliged to learn the corresponding date for each section in different years. In Egypt, peasants sometimes shift from one calendar to the other, or rather use a strange combination of all the three calendars in referring to certain events that happened in the past. Thus a man may be referred to as being born in Ramadan (Islamic) of the year 1952, or even in the year of Nasser's Revolution, which is reckoned by the Western calendar. Such ways of time-reckoning, though confusing, do not in fact cause much inconvenience to the peasants themselves, since the exact dates are of no real importance to them in their routine daily life. Many people, especially of the older generation, do not know the exact date of their birth or marriage; or indeed the year of such events. Instead, they tend to relate them to some other more significant incident which they remember. Time seems then to be, to some extent at least, an order of events of greater significance to the community than to any particular individual. For in the course of time, exact dates, and even the exact year, may be forgotten, and what remains is these outstanding events. Nevertheless, the ordinary uneducated peasant finds himself at a real loss when he tries, in his dealings with government or administrative authorities, to find the corresponding section of the Western calendar for events which he used to reckon according to the traditional calendars.

Responses

All this raises questions that need to be investigated more deeply if we are to reach a better understanding of peasant attitudes towards time. The first question rises from the peasant's obvious lack of interest in organizing activities according to well-defined time units, or in scheduling work to a strict timetable. Are they conscious of this passive attitude towards the time factor, and if so, how do they explain their attitude and justify it to themselves?

There is little doubt that the attitude of peasants towards time is a mere response to the slow rhythm of a life that is determined largely by seasonal variations. As far as traditional life is concerned, peasants see nothing wrong in the way they behave, act and work. On the other hand, failure to achieve what one wanted to do is usually attributed to the interference of supernatural forces such as fate, the evil-eye and even witchcraft. Peasants in Islamic societies are great believers in the interference of God's Will in all they may do or say. They often quote a passage in the Koran that states, explicitly, that no one should say he intends to do anything unless he says, in the same breath, that he will do it with the help of God. All man's deeds, successes and failures are predetermined for him, and this certainly provides the peasant with an excellent and ready excuse for failing to carry out his responsibilities, or at least a ready justification for this failure. A man fails to finish his work in time, not because of negligence or carelessness or laziness or bad planning and miscalculation on his part, but rather because God the Almighty does not want him to do it, or because time is not 'ripe enough' for the work to achieve its preconceived result. As cultivation is determined by specific seasons, and as each crop ripens also at a certain time, so also with human deeds and actions. Nothing can be achieved except at the right season, regardless of the efforts that may be made and the time that may be spent. The right time or the right season is determined by God or by fate. One has only to do all that one can, then leave the results to be determined by fate. Fate may be predictable, but it cannot be controlled.

The second question is concerned with the rather contradictory attitudes of the peasants towards time, its importance and its value. These contradictory attitudes are expressed very explicitly in their proverbs and sayings. They become even clearer when one compares these proverbs with the people's actual behaviour. Proverbs like 'Never postpone to the morrow what can be done today', or 'Time is as precious as gold', or 'Time is as sharp as a sword and will cut you unless you cut it', can be taken as indicators of promptness, exactitude, high evaluation of time and initiative. But proverbs like 'Why hurry! The world will be always there. It will not fly away' give a sense of idleness, hesitancy and inaptitude. The Algerian proverb that says 'Live as if you were to live forever, live as if you were to die tomorrow' illustrates the contradiction that, as Bourdieu says, simultaneously exalts foresight and submission to the passage of time. In Egypt, the people like to quote the Prophet's tradition, 'Do for your actual life in this world as if you will never die; do for the hereafter as if you were to die tomorrow'. This 'tradition' refers to the two worlds: the secular and the religious, to which one belongs and to which one should give equal consideration. This contradiction is in fact a contradiction between the ideal and the actual reality. It is an ideal to be prompt, alert, exact and decisive, and to finish things at the right time. But in daily life peasants tend to be rather slow and to leave for the morrow whatever work they can postpone.

This brings us to the third question, i.e. the bearing of the past and the present on the future. There is a general assumption among many sociologists and anthropologists that peasants are very attached to the present, and that they tend to live in the present moment. Actually, the past has always a fantastic glamour in their eyes and they look at it with nostalgic admiration. The best days are those that have passed and gone; not least because one has already lived them and has known everything about them. The future is void and illusive. One knows nothing about it, and therefore it is usually looked at with scepticism and uncertainty. The future does not belong to man; it belongs to God, and thus one can never tell what may befall one in the future.

But in spite of this scepticism peasants are almost sure that what took place in the past will happen in the future, unless some unexpected factor interferes and changes the ordinary course of things. To the peasant mind, such interference is most unlikely to occur. Peasants are usually regarded as being conservative, perhaps to the extent of rigidity. They dare not venture on new or untrodden tracks. This conservatism may have been the natural outcome of very long dependence in organizing their lives and their activities, especially agricultural activities, on the ecological cycle, which repeats itself with such regularity that one can expect what will happen at a certain season after so many years. The daily life of the peasants is regulated by the regular movement of the sun. Agricultural activities are organized according to the regular seasonal variations, which themselves are determined by the regular motions of certain heavenly bodies. Even in some desert cases in the Middle East where the irrigation of fields depends on subterranean water that flows incessantly from the artesian wells, with the result that irrigation has to take place at night as well as during daytime, the peasants depend on the regular movements of the stars in organizing the distribution of water at night among the fields according to an intricate system of water tenure. In many cultures the appearance of certain stars has been taken as an indicator to determine the proper season for certain activities. The Pleiades are of particular importance in reckoning the seasons and in arranging cultivation and other activities. All this means, in the end, that the regular rhythm of the ecological cycle makes the peasants sure that the past will repeat itself with very minor variations, and that the future is nothing more than another image of the past.

The peasant future

The concept of time as something that recurs and repeats itself, and the inclination to see the future as another image of the past, raise some doubts about the ability, or even the willingness, of the peasant to plan for the future. The crucial question that imposes itself here is: If the future is so illusive and belongs entirely to God, and if man cannot take hold of the future, then to what extent can a peasant make a real and solid plan for the future?

Although sociologists and anthropologists who have dealt with this subject agree that peasants do make certain predictions about the future and arrange their life accordingly, they express very strong doubts about their ability to make far-reaching plans. Bailey, for example, says that peasants hardly think in terms of five-year plans, if that is what is meant by planning. He goes to the extent of arguing that 'the idea of planning can exist only in those cognitive maps which include the idea of man in control of predictable and controllable impersonal forces'. And since peasants give much weight to fate and to other supernatural or mystic forces, which they think do intervene in their actions, then 'policy-making and planning are not part of their cognitive map of the world and human society. They do not reject the idea of planning as wicked; they simply do not have the category'. It is true that peasants do plan the use of their resources, in the sense that they breed their cattle, or save for their marriages, or keep seed for next year's sowing, and so on. But these activities are not, so argues Bailey, planning in the way in which this word is used when one speaks about 'five-year plans'. Moreover, the policy-maker sees a future that is different from the present because it will then be of a completely different kind from what it is now. This is something that falls far beyond the imagination of the peasant, who it seems cannot be interested in figuring a future state of affairs that is radically different from the present.

Bourdieu argues that traditional Algerian society does not try to transform the world but to transform its own attitude to the world. Indeed, he notices the absence of all effort to forecast the future and to master it, and he attributes this to the people's distrust of the future and to their feeling that they cannot take possession of it. The peasant usually refuses to sacrifice a tangible interest to an abstract one 'which cannot be comprehended by concrete intuition'. Thus he cannot sacrifice, for example, the pasture assured to his herds by the land to a mere abstract plan of agricultural development of which he does not know the outcome. Therefore, planning creates no incentive for him, because the plan is based on abstract calculation—which belongs to the realm of possibilities. The peasant certainly has foresight, but does not practise calculated planning; and between the two 'lies the same gulf as between mathematical demonstration and practical demonstration by cutting and folding'.

This scepticism about the peasant's capacity for planning and his interest in it comes from the very concept of time adopted by the peasants themselves. Most social scientists tend in their analysis to distinguish between two main concepts of time, which have been simply called, by Evans-Pritchard, ecological and structural. Ecological time is rather cyclical, as opposed to the structural progressive time. Other social scientists, such as Bailey and Bourdieu, whom we have quoted here, distinguish between what they call 'the round of time' and 'time's arrow' or 'time as an arrow'. The peasant plans for the round of time, in the sense that he allocates resources as if he holds the assumption that the coming year will be 'this year over again', with very minor changes and trivial variations. Planning in the sense of making five-year plans envisages time as an arrow, envisages work as having a beginning and an end, and assumes that there is always a well-defined and clear target to be reached and an aim to be achieved. In other words, the plan for a future that is radically different from the present needs rational and deliberate calculation, thus falling beyond the scope of the peasant's mind or interest. Those who think in terms of the 'round of time' tend to see change as a result of the intervention of mystical and supernatural forces. This intervention makes it vain and hopeless for human beings to plan; it is useless to plan in the face of such forces.

Modern change

The above account is rather formalized and over-formalized. Certain changes are now taking place in peasant attitude towards time and making use of it, in their traditional ways of organizing their activities as well as in their outlook to work. These changes have been taking place for some time as a result of a number of factors. Here the case of Egypt may serve as a good example.

The spread of education, in most developing countries, has brought new ideas and new concepts of time to the young villagers, who have become more acquainted with the modern systems of time-reckoning and with the Western calendar. As a result of introducing compulsory free education in Egypt, the Coptic calendar is now falling into disuse, and the young generation of villagers is looking at it as a mere part of ineffective folk culture. The Islamic calendar is still remembered, but only for its role in regulating major religious occasions. In fact, the young villagers seldom mention any of the Coptic months in their daily parlance, and they hardly know any of the Islamic months except those that have exceptional religious significance. The Coptic calendar, in particular, is giving way to the Western calendar, and this process is taking place very smoothly and in a rather unobtrusive way, mainly because the two calendars are solar. The rhymed phrases that were used to help memorize the Coptic months have been modified to fit the Western calendar. In the Egyptian Western Desert, for example, one may come across such rhymed phrases, which indicate the importance of the various months in the Western calendar for agriculture. One of these phrases refers to March as the most important month in determining the state of the crop (barley), according to whether it rains during this month or does not. We have already referred to what the Algerian peasants say with regard to 'thundering' in January and February.

Agricultural extension programmes have also helped to promote this change. These programmes are usually arranged by the Ministry of Agriculture and carried out by agricultural officers who hold university degrees in agriculture. Having been working for years in rural areas, they know how to convey to the peasants, not only ideas about the importance of new techniques, but also new ideas about the importance of utilizing time effectively and efficiently, the benefits of being punctual in performing one's work and keeping to schedule. This does not mean that peasants are more ready now to use smaller time-units, or keep to the clock in their activities. Far from it. It only means that they are more willing to accept advice about the importance of keeping to strict dates in agricultural work. Sowing, for example, is now carried out at certain specific dates, and not simply according to weather conditions. Precision and punctuality are finding their way, though slowly, into the peasant's life and actions.

One need not mention the role of the mass media, especially the radio, in introducing such new ideas and attitudes. The fact that peasants listen regularly to certain programmes, such as the Koran Recital, the news and the weather bulletin, which are broadcast daily at certain hours, has aroused in them a deep sense of appreciation and a respect for punctuality and precision. It is noteworthy that peasants often ask those who have watches about the 'exact' time, although this knowledge will hardly be used in organizing their activities. It is, nevertheless, a symptom of change and a sign of a new understanding of the role of 'exact' time in life, even though this role is not quite clear yet. In any case, increasing numbers of villagers own watches and use them to organize their activities, especially when it comes to dealing with government offices.

Industrialization is perhaps the most important and most effective factor in changing the traditional concept of time and introducing a completely new outlook towards the role of time in daily life. Although relatively few villagers work as unskilled labour in the new industries and factories that have been established in the rural areas, yet the mere fact that they have to follow a strict timetable in their work has affected, to some extent at least, the life of their families, who have to adjust to the new situation (e.g. getting up in the morning at a certain hour, having meals at a certain time, and so on.

Important as these changes may be, they seem to be somewhat superficial. Industrialization, for example, has not fully changed the modes of thought of the peasants with regard to time. The worker who keeps to schedule in his work hardly follows any timetable in performing his other activities once he leaves the factory after working hours. He goes back to the traditional life, where time has little value and is not taken so seriously. The same can be said about education. For although it has succeeded in making villagers aware of the right scientific significance and interpretation of things, it has not yet succeeded in erasing completely all beliefs in the supernatural forces people think affect their lives, their present and their future.

THE CONCEPT OF TIME
IN RURAL SOCIETIES

Mübeccel B. Kiray

Introduction

Time is a dimension of man's experience. However, this dimension does not correspond to a simple physical reality. The concept of time is 'learned' by our experience of change. The physicist and the biologist must introduce a parameter t to account for the evolution of natural phenomena.

Analysis of our experience and the requirements of science show that two fundamental aspects must be distinguished in the concept of time: (a) the sequence and, more precisely, the order of changes, and (b) the duration of the changes or of periods between them. The interlocking of order and duration defines the process of the change. We experience a great number of series of changes that are apparently autonomous: time of the seasons, time of day and night, time of human life, and so forth.

One way of knowing time is defined by studying temporal behaviour, i.e. by our adaptations, at first, to the sequence and duration of changes, and later, to the multiple, changing series in which we live. These adaptations take place on two levels. The first is common to animal and man. Through learning, activities become synchronous with the speed of changes. The second is peculiar to man, who is able to symbolize the various aspects of the changes. Society teaches these symbols to each one of its members, who uses them to represent the changes to himself, to orient himself within them and also to control them.

Man experiences many changes simultaneously, some of them being periodic. These serve as points of reference, by means of which the others can be located. The nychthemeral rhythm is the most important but not the only one.

From the moment of their birth, organisms are subject to changes and in particular to their own internal changes. The organism adapts itself to the sequence of needs and the means of satisfying them. To the physical sequence there corresponds a perceived sequence. The entire rhythm of the organism (the rhythm of alimentary activity, of sleep, of the body temperature and of all physiological functioning) has a cycle of twenty-four hours. This rhythmic activity turns our organism into a regular clock and gives us points of reference by which we can orient ourselves to the time of the day. Man spontaneously makes use of the temporal points of reference provided by his organism; but in addition to this he learns to orient himself to other periodic changes, the most important of which are provided by the changes of nature (days, years) and the basic social structure of the society to which he belongs.

Man learns to employ socially provided points of reference such as clocks, calendars and so forth. The principle is always the same: to make the experienced moment correspond to the phase of a periodic change that serves as a system of reference.

In addition, man is able to distinguish the present moment from what has been (the past) and what will be (the future). Past and future are made more precise by the learning of the society's language. Along with the language, society transmits its representation of the past to the future. Each social framework (family, type of work, belief system, political organization) in any society has its own way of seeing time and, as Gurvitch clearly demonstrated, we may speak of the multiplicity of social time.[1] In every society, people use different 'times' for different activities. This is particularly striking in transitional periods, when different degrees of precision in dividing time are used in aspects of life where speed is greater because of technological advancement.

In a given society, the temporal horizon appears to be fairly closely bound up with the cycle of experienced expectations and satisfactions. Every man has the capacity to evoke the very distant past or future, but in practice the horizon that has solidity and reality for him is linked to his way of life in his society and culture. The time of a rural population is very different from the time of office workers in the city. It has been found, for example, that the children of farmers make up stories covering shorter and less precise periods of time than the children of the urban middle class.[2]

It should also be kept in mind that every man belonging to several social groups in the same culture has multiple temporal perspectives. He has to pass from one to another—from family time to office time—and try to bring them into accord. Individuals in changing societies organize themselves with different reference systems of time for various activities—particularly, as we shall see below, peasants who become workers or cash-cropping farmers and acquire a new awareness of speed and the need for punctuality.

Historical perspective

Since all human activities occur in time, the existence of social systems necessitates some organization of time. In every society such organization entails: (a) systems of time measurement, based upon cosmic and human cycles; (b) the allocation and scheduling of time by individuals; and (c) a set of attitudes towards time past, time present and time future.

While many thinkers have been interested in the basic categories of the understanding—time, distance, size and so on—it was Durkheim's discussion[3] of the subject that laid the basis for modern social scientific treatments. It was Durkheim who first argued that these categories are not given *a priori* but are social constructs. His observations encouraged other social scientists to study the way in which cultural variations in concepts such as time and space are related to other aspects of social life.

As was said above, the experience of time takes two major forms: sequence and duration. From the standpoint of sequence, events are seen as located in a particular order along a moving continuum. The experience of duration derives from the relative span of events and the intervals between them. Although all societies have some system of time-reckoning, and some idea of sequence and duration, the mode of reckoning clearly varies with the economy, ecology, technical equipment, ritual system and political organization and, most importantly, with changes in these. A peasant system has little need of elaborate scheduling, nor does it always possess the mechanical devices that permit accurate measurement. In non-industrial societies the repetitive patterns of human life and the world of nature provide the basic measures of time-reckoning. These measures can be considered in terms of two main cycles: the human and the cosmic. In agricultural societies, in each cycle, the main points of significant change are marked by rites of passage.

Sequence and duration, cyclical or linear patterns, and systems of reckoning occur in all societies, but the point of emphasis changes. The measurement of long durations of time did not begin before the surplus of agricultural produce reached a significant level and brought the urban revolution, with which came writing and the possibility of elaborate calculations concerning the movements of the heavens.[4] Strict chronology began with the establishment by the Chaldeans of a fixed era, the era of Nabonassar in 747 B.C.[5] The introduction of such a base point for the calculation of years was essential to the arrangement and prediction of long term periodic phenomenon such as taxing of land produce, the basic form of wealth of that society. It was also important for science, e.g. in connection with eclipses. All time measurements up to units of a minute were observed. The need was felt and organized when agriculture became supreme in society, and a surplus was achieved to provide for the non-agricultural population, who in turn devoted considerable time to increasing agricultural production further. Days, months and divisions of the day into five or seven sections were established. Since 'winter' was a quiet period and the 'leisure class' did not require a leisure period on a yearly basis, rest days once a week became usual.

Further refinement of the time concept came with the dominance of non-agricultural, that is industrial, production, with the advance of science and technology, with larger urban settlements, and so on, but we shall not be concerned in this paper with time in industrial society.

Measurement of periods of time

In early agricultural societies, the passage of time was calculated by reference to a series of repetitive units that were measured with varying degrees of precision. Certain of these units were based on the movements of nature—the daily rotation of the earth, the regular phases of the moon, and the annual movement of the earth around the sun. The reckoning of days, months and years occurs universally in pre-industrial peasant societies. But such units are not necessarily organized into an interlocking series with one unit representing a specific fraction (or multiple) of another; instead they may constitute a set of discontinuous time indications.

DAY AND NIGHT

In all societies, and particularly in agricultural ones, some division of the day is made according to the position of the sun in the sky. Hence concepts of dawn, forenoon, afternoon and sunset, usually expressed with the activities peculiar to that time of the day, appear universally. In central Anatolian villages in 1944, ten divisions of the day were used, such as first rooster, dawn, 'leaving of oxen' (for grazing), sunrise, mid-morning, noon, 'return of oxen', sunset, evening and midnight.[6] Freqeuntly the periods of dawn and dusk, the times that call for a reorientation of activities from those of night to those of day, are further subdivided, and the terminology is refined.

The division of life into day-light and night-dark, movement and rest, waking and sleeping often provides a symbolic framework for many other social activities. Night is generally seen as linked with evil, with witchcraft and with illicit behaviour of all kinds. It is a time for supernatural agencies to reveal themselves in dreams and for spirits of various shapes and sizes to roam the earth. Night is also the time for sleep and love. The daytime belongs to the productive activities, and there is a lack of activity outside the house at night.

With the use of the sundial the variation in the position of shadows was formalized. Time-reckoning thus moved in the direction of regular divisions of the night and day into seconds, minutes and hours, a systematization that runs counter to the ideas of the peasants in lands where there are seasonal differences in the duration of daylight. Daylight, of course, was very important in the life of the peasantry. However, clocks were devised in agricultural societies from Babylonia and Egypt to China, wherever control of the agricultural wealth, its distribution and transportation meant precise recording not only of years and seasons but also of days and their fractions. The precise divisions made by the clock are essential to any elaborate system of interaction that relates to the economic, military, political and ritual needs of agricultural societies. It seems that the precision achieved by the 'scientist' of these societies in time-reckoning was as remote from and unrelated to the standards used by the ordinary peasant as is the precision achieved by the pure scientist from the standards used by the man in the street today.

On the other hand, the methods of time-keeping seem also to have been stimulated by magical and religious concerns as much as by pragmatic interest in peasant-agricultural societies. The measurement of hours by sundials and by sand-glasses was an ecclesiastical demand and so too were ideas of punctuality. Members of the regular clergy of the mediaeval monastery were enjoined to organize their lives 'by rule', that is, by a specific allocation of time for work, for sleep and for worship. The ringing of the prayer bell seven times a day established time by means of instruments that divided the day into regular intervals.

Other world religions developed their own diurnal ordering of time. Followers of Islam, for example, are required to offer the five canonical prayers (*salat*) at fixed times during the course of the day.

THE WEEK

The week has no definite basis in the external environment. It is an entirely social construct, basically in agricultural societies, varying in length from society to society—seven days in the Judaeo-Christian-Islam world, three, four, five, or six days in certain part of West Africa, South-East Asia and Central America. In early Rome it was eight days. In China it was ten. However, the weekly cycle always consists of a relatively small number of days (usually named) and is used to regulate short-term, recurrent activities, especially those of the market place where limited agricultural produce is exchanged. In Mesopotamia the seven-day period was linked to the five planets together with the sun and the moon in a planetary or astrological week. The seven-day week spread through Europe, North Africa, India and the Malay peninsula to all peasant societies. It is used more or less universally today, though present names of the days continue to indicate the pre-Christian origin.

The LoDogao, a primarily agricultural society of northern Ghana, designates each of the six days of the week by the name of the village where a market takes place on the day in question.[7] The very terms for 'day' and 'market' are the same (*daa*) and the weekly cycle is simply *daar*, 'a plurality of markets', so that the names of the days not only record the pattern of market gatherings but serve as a measuring rod for other short-range activities.

The importance of market time is again illustrated in early mediaeval England, where each neighbouring town held its market or cheaping (hence cheap) on a different day of the week. The inhabitants of outlying districts would come into trade and also to have an opportunity of meeting together so that disputes could be settled, marriages arranged, and social contact enjoyed. Thus in almost all peasant societies the market week is a way of organizing economic exchange as well as other social time. A similar grouping of days has been observed in isolated villages of Turkey as well.

The weekly cycle of markets differentiates one day from another and serves to break up agricultural activities by providing some change of pace, substituting rest for work, exchange for production. In general in agricultural societies there is also a weekly shift from the profane to the sacred, since a special day is allocated for religious activity. In Islam this day is Friday, in Judaism Saturday and in Christianity Sunday. These calendrical differences reflect distinctions of theology and organization.

The polytheistic religions of agricultural West Africa also have their day of rest, which coincides with the market day. It has been suggested that in West Africa, while the basis of the week is economic, the rest day is religious in origin. The days are usually named after markets, but rest days after the gods worshipped on these days. Among the LoDogao, one day each week is set aside as a 'day of not using the hoe' when iron implements are forbidden. It is on this day that important sacrifices are made to the earth shrine, under whose aegis all major uses of the soil, house-building, farming, burial and ironwork are undertaken.[8]

THE MONTH AND THE YEAR

While itself consisting of a specified number of days, the week is rarely a subdivision of larger units of time measurement. The next unit in size, and one that is given universal recognition, in some form or other, is the month, based upon the lunar cycle of 29.5 days. Unlike weeks, months are usually thought of as organized segments of a seasonal cycle. All societies recognize some kind of yearly cycle, since this is required by both agriculture and hunting. Agriculture in particular demands an annual scheduling that determines the allocation of work and food—which is harvested once in twelve months—as well as the setting aside of seed at harvest time to be preserved until the next planting season. No agricultural society can avoid some long-term budgeting of this kind. The tropical paradise where wild fruits offer a natural superabundance of food and drink is a figment of the imagination of industrial urban environments. Nobody in fact just passes the time, although it often seems so to those dependent upon more demanding schedules. There are, of course, outstanding differences in the degree of accuracy required by different schedules.

While the weekly markets of mediaeval England catered for the local trade, there also existed the yearly fairs or markets to which traders came from all around. The tolls of these fairs were often allocated to various ecclesiastical foundations, and the specific day on which the market took place was sometimes the saint's day of the religious house, so that the fair doubled as a fete and the traders as pilgrims. Similar fairs are usual in all peasant societies. An activity of this kind that brings people together from widely separated places at a specific time of the year clearly requires a more accurate calculation than is provided by a simple count of moons, loosely linked to a seasonal cycle. It demands a calendar or natural occurrence that is accurate, regular and widely known, so that precise co-ordination on an annual basis is possible.

One difficulty in constructing such a system lies in the fact that no sum of lunar months adds up to a yearly cycle. Intercalation is necessary in order to reconcile the year of twelve lunar months with the solar year on which crops depend. In the usual practice, no fixed number of days is assigned to the lunar month (in isolated Islamic communities, for instance, it begins when the new moon is seen), and likewise the year is considered to begin when the appropriate season comes round, the length of the months being adjusted accordingly. Thus the harvest moon comes when the harvest is ready, and the planting moon is set by some biological clock, some natural phenomena or 'watch of flower'. The poor peasants of the Tauros mountains, for example, start to travel down to the plains to pick cotton when thistles are all dried and dead.[9]

The abandonment of the lunar month results in a non-lunar month, or 'mesne', under which term can be included any unit greater than ten days and less than a year. The Ashanti in West Africa have the *adae*, a period of forty-two days, formed by the intersection of a six-day and seven-day weekly cycle, the first of local origin, the second probably a Muslim derivation.[10] Both the greater *adae*, which occurs after eighteen days, and the lesser *adae*, which takes place after a further twenty-four days, were important occasions for sacrifices to the royal ancestors. The *adae* (although unnamed) provides a continuous chain type of calendar, divorced from the seasonal cycle (as West African seasons were not 'pronounced') but linked to a complex series of politico-religious festivals.

The necessity for a closer 'fit' between lunar months and solar years came only with the introduction of written calendars, which eventually led to the abandonment of the lunar month, as in the Julian calendar. Historically, the first breakthrough towards the modern Western system appears to have been made in Egypt, which established a year of twelve non-lunar months each with thirty days. This is a country where agriculture does not depend on four seasons but on the periodicity of the flow of the Nile River.

The formalization and elaboration of written calendars took place under a variety of pressures: the demands of agricultural planning, particularly in complex irrigation societies, and the organization of trade, especially long-distance trade in the Mediterranean basin, where the agricultural wealth was regularly transported to the capitals of the empires of various eras from producing societies all around the basin. It was also important for the military and administrative activities of such political organizations.

One factor that inhibited navigation was the existence of purely local systems of time-reckoning in agricultural societies. The universal standardization and co-ordination of time belts had to wait for new experiences in speed and distance, such as those provided by transcontinental railroads in the United States around 1880.

With the development of writing, astronomy and mechanics, there occurs an increasing dissociation of time measurement from the commonplace events of rural societies—the growth of crops, the movements of animals and the human activities to which they are directly linked. Systems of time-reckoning have thus evolved from concrete time indications of a discontinuous kind, with increasingly abstract, numerical and regular division linked to a continuous calendar based on the movements in industrial urban societies. However, the development of more abstract time-scales tends to supplement rather than replace more concrete concepts of time for various aspects of social life and to enable us to speak about different social times for many subjective experiences in every social system. The villagers and farmers of modern industrial society still constitute the group that lives with the least demanding schedule and the least precise time standards in comparison with its urban middle and upper classes.[11]

Social change and the concept of time

In a study made in 1962 we attempted to analyse the resultant re-standardization of the concept of time in various fast-changing rural communities in northern Turkey.[12] There the farming was in the process of change from self-sufficiency to cash-cropping. Efficient road and motor transportation had been established, with new dimensions of mobility and speed in space and new aspects of social structure.

It was thought interesting to make a survey of systematic samples to show the degree of change and variations in the concept of time, and the diversity of such concepts in the life of the individual. In such a society different people would be using different anchorages in structuring their perception of time, and there would be many social times for individuals in various aspects of their lives.

Following such a hypothesis, we asked the heads of families at what hours they got up in the morning. Here was a question on a most common activity, but very much related to the style of life of the respondent. Basic activities such as the mode of production, family organization or system of belief would all affect the time of the individual's rising in the morning and its conceptualization. As will be seen in Table 1, in the town the ratio of universal standard time has become dominant (86.8 per cent). But in its surrounding villages the same ratio falls down to 61.7 per cent. However, even in town, people who organize their daily life according to the periodicity of nature or praying time had not yet disappeared altogether. It is also interesting to note that those who use concepts other than international standard time may change their social time when travelling by bus. The duration of a trip to the town is again expressed in terms of international standard units by more than 91 per cent of the respondents. When this is compared with the dominant concepts used for the durations of similar trips (in Sherif and Sherif, 1944), evidenced in remarks such as 'you start early in the morning and reach there by sunset', the scale of change is particularly noticeable.

In the town and in surrounding villages, our research showed that even the population of the most isolated villages with new experiences, particularly in speed, had started to use international units in one or more aspects of their perception of time. Proof of such behaviour is the frequency of having mechanical devices for measuring standardized concept of time, i.e., clocks and watches. In the urban community studied, 95.7 per cent of the household heads had proper clocks or watches and 82.4 per cent had calendars. The ratio for villages was only 82.3 per cent for clocks and watches, 63.2 per cent for calendars. Although lower than for the town, it was much higher than for villages in 1944 (in Sherif and Sherif) and unexpected for an agricultural environment that was establishing its first contacts with the wider world. On the basis of the findings of Sherif in 1944 and Kiray in 1964, as a result of differential contact with modern technology and different intensity of integration in more differentiated and organized social environments, it is possible to conclude that when individuals in a rural social organization have little contact with modern technology and the internationally standardized units of time that accompany it, their 'standardization' (their anchorage) depends on (a) periodicity of work activities and socio-economic activities (like market days) or (b) the periodicity of natural events (like sunrise and sunset, or cycles of the moon). The anchorages standardized on these bases lack precision in varying degrees.[13] But as one passes from more isolated to less isolated communities, or as the community becomes more developed, international units of time are used roughly in proportion to the degree of the impact of modern technology and the experience of speed, and their use becomes correspondingly more precise.

TABLE 1. Concepts of time in Turkish communities

Concept used	Rising in the morning		Trip to a certain city	
	Town %	Villages %	Town %	Villages %
Standard international time unit	86.8	61.7	92.0	91.2
Periodicity of nature	5.0	26.5	0.6	2.9
Praying time	5.4	10.3	1.0	—
Not clear	2.8	1.5	6.4	5.9
TOTAL	100.0	100.0	100.0	100.0
$N =$	484	68	484	68

Source: M. Kiray, *Eregli: Agir Sanayiden Once Bir Sahil Kasabasi* [Eregli, A Coastal Town Before Heavy Industry], Ankara, Devlet Planlama Teskilati, 1964.

It also seems that the greater the degree of contact with modern means of transportation and other technological products, the greater also is the actual mobility of individuals and the greater their psychological mobility. People who have had less contact with modern technology and long-established relatedness to familiar surroundings have less psychological mobility.

Attitudes towards time

The norms, values and attitudes about various aspects of time also change in each culture. The attitudes about past, present and future on the one hand, precision and punctuality on the other hand, acquire different meanings and emphases according to the complexity of the society and the speed and rhythm experienced by the technology achieved in that society.

In non-literate cultures, and among the peasants of great civilizations, ideas and attitudes concerning the past tend to reflect present concerns. There the past is a backward projection of the present. It is only when writing gives a material embodiment to speech that the distant past can represent more than a backward extension of the present.

Literacy influences attitudes not only towards the past but towards the present and future as well. The permanence of written records makes a radical difference in the accumulation and storage of knowledge, and opinions thus create the possibility of more rapid change.

But above all, attitudes towards present-time focus upon the alternative uses of time that are offered by an elaborate division of labour, the minute scheduling that the division of labour entails, and the continued presence of the watch upon the wrist that makes man ever conscious of the fleeting moment. In industrial societies man is made aware that time is the scarcest resource; he learns to 'speed' and 'save' it like money. The rural man is not used to 'watching the clock' or to the demanding routine of much of factory life. For the peasant, and even the farmer of modern industrial society, time has been nature's time, and the organization of his activities largely his own affair. However, the ex-peasant of transitional societies, for instance in the Mediterranean basin, is today adjusting himself to the industrial order of time in the factories of the north-western Europe. In general he seems to be adjusted to the rhythm of work by watching the clock. However, when he returns for vacation or goes home permanently, he does not fall back exactly to the old pattern but with his watch on his wrist he still lives a much less structured schedule.[14]

The factors that bear upon attitudes towards the present are also relevant to attitudes towards the future; indeed, the difference simply turns on the question of scheduling over the longer term, as compared to the shorter term. In peasant communities the scheduling of production operates on an annual basis, but it is largely repetitive and future activity is mainly a continuation of the present. However, in today's fast-changing rural areas, future orientation with planning for new activities and cash-cropping has brought to the population new attitudes towards the future. In a study comparing social change in four villages on different levels of technological development in southern Turkey it was observed that the higher the level of development, the more confidence and hope for the future prevailed.[15] Still, even the least-developed village could not be seen as past-oriented and fatalistic to such a degree as to neglect planning for the future (Table 2). In fact it seems that once change sets in in the vicinity of any settlement, positive attitudes towards change, time and the future are the first to be accepted among the peasants. The possibility of long-range planning is vastly increased by the existence and functional usage of writing. Plans, projected organization of future time, are as much a requisite of the personal sphere as of the national and industrial domains. The changing peasant also brings planning to his life, with future projects for investment and cash-cropping, in a striking fashion. In the four Turkish villages it was also observed that the great majority of households had definite plans for the future that would make them really farmers, the agricultural population of an industrial society and no longer peasants.[16] Thus one of the peasant attitudes towards time that is held partly responsible for lack of change and development—namely lack of planning and future orientation no longer prevails. It seems that after change starts the rest comes quickly—so much so that the ex-peasants who have been set free from the land, because of mechanization and other modes of development, but cannot be absorbed into industrial and organizational jobs in their society's urban centres with equal facility, are forced to migrate to highly industrialized, labour-demanding north-western Europe. There they adjust to the requirements of more precise time and speed with remarkable ease.

In most societies the distant past and distant future tend to be of peripheral interest. By and large, the 'other world' of agricultural societies is visualized as a continuation of this world, although there is usually some distribution of rewards and punishments for behaviour on earth. And when a culture of the less complex kind is hard pressed, typically by contact with industrial urban society, there is a tendency to seek comfort in ritual designed to bend time backward to an earlier paradise, or to advance time and hurry on the advent of the Messiah or the coming of the millennium.[17] In more complex situations, in the process of de-peasantization of rural population, the larger political system of the society creates parties with new re-interpretations of systems of belief, distant future and after-life, which become successful only for a short while. In the increasing pace of social change and individual mobility in industrial urban societies and in its farming population, millennial dreams are replaced by political utopias, the idea of a fixed destiny by a concern with educational mobility, and the belief in immortality by a concept of social progress and continuity.

TABLE 2. Wishes and plans for the future in selected Turkish villages

Wishes and/or plans	Least developed %	Medium %	Most developed %
Higher standard of living	47.1	54.2	45.8
Increased production	20.6	27.1	38.1
More education	8.8	6.2	3.8
Move to city	5.9	—	0.9
Fatalistic answers	17.6	10.4	9.5
Not clear	—	2.1	1.9
TOTAL	100.0	100.0	100.0

Source: M. Kiray and J. Hinderink, *Social Stratification as an Obstacle to Development*, p. 217, Table 34, New York, Praeger, 1970.

In any case it is difficult, if not impossible, for man in general and for rural populations in particular, to envisage a final end to time, on either a cosmic or an individual level, and the continuous claim of family living helps to mitigate the prospect of complete finality.

NOTES

1. Georges Gurvitch, 'Structures Sociales et Multiplicité des Temps, Société Française de Philosophie, *Bulletin* 99, 1958, p. 142.
2. Lawrence L. Leshan, 'Time Orientation and Social Class', *Journal of Abnormal and Social Psychology*, Vol. 47, 1952, p. 589–92.
3. Emile Durkheim, *The Elementary Forms of the Religious Life*, London, Allen and Muroin, 1954 (1912).
4. Gordon Childe, *What Happened in History*. Harmondsworth, Penguin, 1964 (1940).
5. Franz Cumont, *Astrology and Religion Among the Greeks and Romans*, New York, Putnam, 1912.
6. M. Sherif and C. Sherif, *An Outline of Social Psychology*, New York, Harper, 1956.
7. Thomas Northcote, 'The Week in West Africa', *Journal of the Royal Anthropological Institute of Great Britain and Ireland*, Vol. 54, p. 183–209.
8. ibid.
9. Yashar Kemal, *Iron Earth and Copper Sky*, London, Collins, 1974.
10. Meyer Fortes, 'Time and social structure: an Ashanti Case Study', in Meyer Fortes (ed.), *Social Structure: Studies Presented to A. R. Radcliffe-Brown*. New York, Russell, 1949.
11. James West, *Plainville U.S.A.*, New York, Columbia University Press, 1945; and Art Gallaher, Jr., *Plainville Fifteen Years Later*, New York, Columbia University Press 1961. See also Leshan, op. cit.
12. Mübeccel Kiray, *Eregli: Agir Sanayiden Once Bir Sahil Kasabasi* [Eregli: A Coastal Town Before Heavy Industry], Ankara, Devlet Planlama Teskilati, 1964.
13. Sherif, op. cit.; Kiray, op. cit.
14. Nermin Abadan-Unat, 'Turkish External Migration and Social Mobility', in P. Benedict, E. Tumertekin and F. Mansur (eds.), *Turkey: Geographic and Social Perspectives*, Leiden, E. J. Brill, 1974.
15. J. Hinderink and M. Kiray, *Social Stratification as an Obstacle to Development*, New York, Praeger, 1970.
16. Abadan-Unat, op. cit.
17. Hinderink, op. cit.

BIBLIOGRAPHY

Bohannan, P. J. Concepts of Time Among the Tiv of Nigeria. *Southwestern Journal of Anthropology*, Vol. 9, p. 251–62.
De Grazia, S. *Of Time, Work and Leisure*. New York, Twentieth Century Fund, 1962.
Fraisse, P. *The Psychology of Time*. New York, Harper, 1963.
Fraser, J. T. (ed.) *The Voices of Time*. New York, George Braziller, 1966.
Hallowell, A. I. Temporal Orientation in Western Civilization and in a Preliterate Society. *American Anthropologist*, New Series, Vol. 39, p. 647–70.

MEYERHOFF, H. *Time in Literature*. Berkeley, University of California Press, 1966.

MOORE, W. E. *Man, Time and Society*. New York, Wiley, 1963.

PIAGET, J. *Le Développement de la Notion de Temps chez L'Enfant*. Paris, Presses Universitaires de France, 1946.

THOMAS, N. W. The Week in West Africa, *Journal of the Royal Anthropological Institute of Great Britain and Ireland*, Vol. 54, 1924, p. 183–209.

THRUPP, S. L. (Ed.). *Millennial Dreams in Action: Essays in Comparative Study*. The Hague, Mouton, 1962. Comparative Studies in Society and History Supplement No. 2.

WHITE, L. *Medieval Technology and Social Change*. Oxford, Clarendon Press, 1962.

THE UNIFYING TIME OF ENVIRONMENTAL TRANSFORMATION

Josefina Mena Abraham

The study of socio-environmental phenomena in recent decades makes us aware of the sharpening contradictions mapped on to the environment of a decaying perception of human nature:

The increasing isolation of geographical zones dealing with production from those dealing with the relationships of consumption and decision-making, i.e. the formation of the industrial sector and the *cité dortoir*. The energy crisis is perceived as a conflict between countries producing and countries consuming petrol.

The ever-widening side-effects of an ill-defined relationship with the environment: pollution, incidence of mental illness, housing problems, production oriented towards destruction, and the concentration of poverty in large zones where the inaccessibility of goods to the majority simply leaves isolated oases of plenty in consumer-cities.

These contradictions acquire their largest expression in cities and entire countries rapidly transformed by the continuous presence of the equation 'popular insurrection–military repression'. We recognize such transformation as a qualitative change. We have no doubt that human consciousness enters a new dimension as we acquire a new perception of human nature.

Recent work on psychology, neurology, and cybernetics has shed some light on such environmental social phenomena. In general terms, this work has, on the one hand, ruled out any possible conception of an 'automatic nervous system' existing independently; on the other hand, it has given evidence on the differential specialization of zones in the human brain with qualitatively different information-processing tasks; and finally, it has shown how perception is constructed by biomental processes. Some of this research has identified a duality between the two cerebral hemispheres,[1] or between the cortex and the striate part (senso-brain).[2] One undertakes a sequential and the other a simultaneous information-processing task.

This paper is an attempt to incorporate some of this work into the environmental sciences in order to show the great importance that timing—temporal experience—has in our ability to transform matter and to transform ourselves. By applying this knowledge to our own social practice, we have become aware of the importance of sensory information to stimulate biomental processes that make possible the creation of concepts, and the transformation of value systems. This has proved of particular importance in organizing communities. Our own concepts of the environment have changed qualitatively.

The very fact that we perceive an environment in movement[3] gives rise to temporal experience. The dialectic feedback from the environment to the individual and back to the environment is fundamentally a biomental process, and it is the only guarantee against the private ownership of knowledge that scientific endeavour essentially rejects.

We advocate a dynamic conception of the cognition process and a dialectic relationship—not a complementary one—between the sequential, analytic mode of perception and the simultaneous, spatial mode. The ability of man to use either mode accordingly for social and political objectives is essentially a problem of evaluation and not of understanding of environmental reality. This ability also means to use time and not to have time, to be a biological clock instead of searching for a separate biological clock somewhere in our bodies.

The internal contradiction in the brain

HOW DO WE CONSTRUCT TRANSFORMATION?

The environment is people, things, animals and trees. There is only one fact about it that matters to us here: it is in continuous transformation.

How do we construct the perception of such transformation? In order to perceive the environment, the brain, stimulated by the perceptual system, makes inferences and tests hypotheses. It is the dialectical interaction between the perceptual system and the senso-brain[4] that selects hypotheses provided by the cortex in a probabilistic manner.

Cerebral zones are able to act upon the others, generating a different power structure, or order, within ourselves to interact with the world around us. The dominance of any of the zones gives rise to qualitatively different bio-mental processes that generate modes of perception. These biomental processes transform sensory information, so enabling us to project socially a certain hypothesis stimulated by the perceptual system. Such dialectical feedback generates our value system, which in turn influences our mode of perception.

Each mode of perception has its own timing and implies qualitatively different temporal and spatial experiences. If we consider that consciousness is organized matter, we can assume that a different consciousness develops with these modes of organizing matter. To see these modes in terms of dualities—cortical/sensorial—is already a limitation of our rationally built language. Keeping this in mind, we consider that the sequential mode operates similarly to a digital computer, and the simultaneous mode to an analogue computer. Therefore, apart from the conventional memory (digital information storage) used to explain and organize data in terms of dualities, we recognize an image-memory that accumulates data in a qualitatively different knowledge. In the latter, the simultaneous processing of many inputs is constantly transforming bits of information by transforming images and overlapping them with others; our sequential—verbal—information processing cannot keep track of the time of such transformations, and cannot express them in terms of dualities.[5]

The generator of our image-memory is the environment. It is the systematization of experience—gained by our intervention in the total environment—that transforms our value system.

A PHOTOGRAPHER IN OUR BRAIN

Our value system intervenes—in our brain's operations—as a photographer rather than as a camera. To illustrate this statement, the experiments of Andreas Adam are of particular importance. The object of his work is

to find methods enabling us to confront systematically different aspects of a representation of reality . . . there is no single representation of architecture which is objective; any one view is only a partial view of the whole . . . every single one, an interpretation of it, representing an attitude to the theme. Postcards representing the city show the development of a certain historical situation; each one represents a phase in a larger historical context. The structural value system which generates a situation is not visible at once through this representation; here the situation is only described as an appearance. A possible method to perceive the context (deep structure) is to confront different phases of the situation; or to confront different attitudes to the theme. Only then, the situation is describable not only as an appearance, but also as something comprehensible. What becomes, then, important is the feedback one gets through the confrontation of a series of pictures.[6]

Through Adam's series we perceive the change in scale the buildings undergo: our visual reference parameters change as a very different pattern appears to our eyes during each timeperiod. We perceive how 'any partial view of reality can conceal the intentions of that one who describes it (the photographer); the instruments that, for instance, a property owner has to conceal his intentions are today increasingly reduced by the unidimensionality of economic development; the building then appears as merchandise with a limited life-span'.

In the perspective of this paper, what is most interesting about Adam's work is to internalize the object-subject relationships that generate these patterns. We have a photographer in our brain. Like a photographer, our value system selects from the whole; the partial views of context and scale thus selected establish the references parameter. Different photographers/value systems create different perceptions of the patterns conditioning the stimulus our brain receives, generating different attitudes in ourselves, that is, different modes of perception.

LOGICAL AND PRELOGICAL MODES OF PERCEPTION

We call these modes of perception logical and prelogical. The important questions for us are: What are the environmental conditions that stimulate one or the other? How does one mode become the dominant one? How does each knowledge emerge? What is our temporal experience of the environment under those conditions?

We know that the logical mode provides us with instruments for analysis and rational thought. But it achieves this by filtering out content and by organizing information so that we give credibility to polarized abstractions; it processes information linearly projecting assumptions in sequential time. This enables us to intervene in the world around us in a detached manner, i.e. we become self-conscious of it. The clock is, therefore, a means of unlinking ourselves from sensorial stimuli. It is our relationship of credibility with the clock that enables us to co-ordinate an action that is timed accordingly with an environment other than the one we are experiencing there and then.

The prelogical mode, owing to its simultaneous processing of many inputs, prevents such detached intervention in the environment. Nevertheless, it gives us instruments to co-ordinate actions that are most probably in tune with people and things around us; this increases the probability of projecting socially the transformation of the sensory information we receive, given that other people receive it as well under the same conditions.[7]

Western society has developed the logical mode above all. Hence its compulsion to control and plan, and its chronological sense of history. This instrument of communication has been used as an instrument of domination over other cultures that retain a prelogical mode of perception, and whose capacity for conceiving abstractions has been very little developed. Persistence in this mode ends by producing an ambience which man cannot any longer integrate into his experience.[8]

Perhaps the first scientific evidence of a conflict between these modes was provided by Freud's work; it documented the appearance of unopenable 'gates' between our cerebral hemispheres. Unfortunately these gates were considered as part of human nature, and not as symptoms of a decaying biomental process.[9]

IDEALISM AND MATERIALISM

In Western philosophy the logical mode finds its theoretical framework in idealism; the prelogical one in materialism.

The conflict between them originates in the antagonism between Plato and Aristotle; it acquires its most modern expression in Hegel[10] and Marx[11] in spite of both having advocated dialectics as a path to the knowledge of reality.

For Hegel, only ideas develop in time, according to the laws of dialectics; that is, they develop gradually—from the simple to the complex—through the development of internal contradictions. To develop internal contradictions is to transform one into the other by taking each other's position.

According to Hegel, natural transformations are determined by ideas: nature develops in space but not in time. This is his point of rupture with Marx and with Einstein, whose theory of relativity established relationships between spatial-temporal properties of matter and the velocity of its movement.

For Marx, matter does intervene actively in these transformations. Marx's concept of the 'qualitative jump' teaches us how—by a dialectical process—we transcend from quantity into quality.

MAN'S PATH TO SOCIAL ACTION

When man discovered fire, he first applied it to his concrete needs (warmth, cooking, etc.). Through practice he consolidated his sensitive knowledge of the technique, and his perception of it changed, moving from the concrete to the abstract. At this point in sequential time man developed the phenomenon, becoming conscious of the relationships between the elements he employed to produce fire (means of production). Through practising the technique socially, man became conscious of the relationships between people employing fire (relationships of production).

The saturation of quantitative modifications brings the phenomenon to the point of qualitative change. At this point, man can decide to use his knowledge as an instrument of power by idealizing fire (making an abstraction of fire into a god, or seeking logical explanations and indulging himself for it as a superior being). Alternatively, he can decide to loop the process and achieve a qualitative change in his concept of fire by changing the mechanical means of producing it to other means—for instance, discovering gunpowder.

COMING TO TERMS WITH BIOMENTAL CONTRADICTIONS

Science has now provided us with instruments that detect, filter, and amplify brain waves and convert them into sound for use as sensory feedback. Variations in loudness help us to learn to discriminate internal states and achieve a desired relationship with our brains. Biofeedback experiments illustrate very clearly how it is possible to control internal muscles and achieve self-regulation.

Conscious awareness of our capacity for self-regulation enables us to demystify both magic and logic. The key question is what to use this ability for.

THE HUICHOL INDIANS' CONCEPT OF ENERGY

The Huichol Indians in Mexico have probably the most authentic indigenous culture still alive in Latin America. They have refused European culture for more than 400 years.

After living with them, the author has come to the conclusion that they have a concept of energy that gives a clear objective to man's ability for self-regulation: that is, communication by developing perception to the level of an associative consciousness. They have a deep knowledge of the conditions which enable a pre-logical mode of perception to emerge; the ability to tune in with this mode enables us to experience what has been called 'our other reality'[12]; this means: consciously to reach the set of images that condition what we perceive. The Huichol Indians are known to have telepathic communication, and the outside observer has the feeling that they know why they have moved up to the top of the mountains, leaving European culture down in the big central valley of Mexico.

Their language has no third person and no future. Children are not private property, and the whole community joins in telepathic orgasms. These cultures, we should not forget, were not influenced by the Cartesian dichotomy between 'the body and the soul'. For them, there is nothing but energy generated by matter in movement through a constant biomental feedback—inwards/outwards—from the individual to his total living environment. The key to the generation of this energy is its constant recycling. We are an integral part of matter, though we are able to become conscious of ourselves and think of matter, i.e. think of ourselves as separate entities in terms of sequential time.

We have all become aligned with such concrete perception of human nature. The problem is reduced to the 'know-how' to organize ourselves with matter: how to develop biologically our thinking cortex, i.e. to become conscious of the movement of matter, to internalize its sequential time, and to experience matter in terms of either mode of perception.

It is therefore clear for the author that both kinds of knowledge—one of visual and one of non-visual characteristics—can interact dialectically and constantly. The conditions enabling such phenomena to occur—whatever they are in a particular situation—are directly related to the concrete reality around us, through direct or indirect experience.

THE DIALECTICAL MODE OF PERCEPTION

The internal contradiction between information organized within these two types of knowledge expresses itself in terms of the fundamental contradiction between society and nature. This contradiction is solved by the dialectical relation of sensitive knowledge to logical knowledge in the development of social practice in production and in the creation of culture, i.e. the development of the forces and relationships of production and reproduction.

But there is no authentic development unless we become rationally conscious of the origins of our value system, gradually transforming—through socio-environmental intervention—its parameters into variables.

We call this internalizing the experience of the external world. The human experiential process is a movement between this internalization (which conditions our internal stability) and the externalization of this process (which determines social action). This movement is not sequential in itself. What is sequential is the interplay between the more and the less organized matter, where the more organized becomes aware that it cannot control the whole phenomenon but can see why.

This movement generates an energy which Western culture has forgotten. We simply indulge ourselves in trying to explain everything, even that which is not needed. This generates a time dephasing: we leave the present and place ourselves in the abstract space of the future, unable to solve the contradiction between the sequential and the simultaneous temporal experiences.

Environmental mapping of opposites

The importance of the environment in stimulating, with sensory information, biomental processes to get over such contradictions becomes clear, as does man's own act of transformation of the environment.

Looking at the village of Kejara (Bororo Indians, Brazil) we realize how environment can incorporate two sets of contradictions and relate them to two different axes. The Bororos have developed a value system based on transcending the opposites by the mutual relation of the contradictions. Their circular villages are divided into two halves: the *Cera* (weak) and the *Tugaré* (strong). A person always belongs to his mother's half and can only marry someone from the other side. Thus a man of the *Cera* area can only marry a woman of the *Tugaré* area; he does it by crossing the *Cera* door if he is *Tugaré* or vice versa. Levi-Strauss has observed that a married man never feels at home in the family house. His house, the one he was born in and the source of his childhood impressions, is on the other side: it is the house of his mother and sisters, now inhabited by their husbands.[13]

A member of one half has rights and duties always to the benefit of or with the help of the other half. The halves are like two football teams that, instead of having a competitive strategy to counter-attack, compete to be perfect in their generosity to each other. This competition to achieve generosity directs man's interests always outside himself; it focuses man's endeavour on to a continuous feedback between man's cortical and senso-brain operations. (A similar system of values exists among some Indians of the south of Mexico.)

A person is in contradiction with another person if seen according to one axis; but by using another axis as a reference parameter they can perceive each in the other's position.

The Salesian missionaries in the region of the Río das Gracas forced the Indians to abandon their village for another in which the huts were laid out in parallel rows. It seems that their objective was to convert the Indians to Christianity. We can analyse this phenomenon from a formalistic point of view, and attribute cultural change in the Indians to the different location of the huts, their shape, size, etc. Or we can analyse it from an idealistic point of view, and perceive the cultural changes as determined by the changes the missionaries imposed on the Indians (the lack of the Indians' participation in the decision-making process).

The structuralist analysis focuses on finding the environmental clues to understand the social structure: 'deprived of the plan which provided an argument to their knowledge, the natives quickly lost their sense of tradition'. The new schema was too simple to sustain the complexity of their social and religious systems.

We see the phenomenon in this way: the village's environmental devices gave clues for understanding the true nature of antagonisms by providing sensory stimuli 'to loop the mental process'; the fundamental contradiction between nature and society was resolved externally by the intervention of the missionaries, and therefore they simply provoked a quantitative change; the missionaries were unable to confront the internal contradictions in the Indian culture because they had a different mode of perception. They introduced an external contradiction: an environment lacking the sensory stimuli that re-vindicated the value system.

The Indians might seem to have lost their traditions, but what we think actually happened was a diminution in the importance of their traditions in their lives due to the overlapping of their value-system with semantically reduced environmental conditions, and a cut in the loop of the biomental process, removing stimuli from the cortex—this has a coarsening effect.

Such an instrument of communication was probably one of the best devices to use to dominate the Latin American Indians.[14] The Europeans could move and communicate in an abstract space; they used very few environmental devices to communicate between them; they had a very detached view of nature and therefore could affect it with actions co-ordinated in sequential time. Their language had a different set of references and logical structure; words could be violent (Indian languages have no insults; the word is water and therefore it is an instrument for providing a flow of waves among people).[15]

In the situation described there was no development. Such an idealistic culture could not find a common point of view within the Indian concrete culture.[16] In this situation we can only reach a state of quantitative saturation.[17]

Time of environmental action

The temporal experience of the dialectical interaction between the perceptual system and the senso-brain is the time of environmental social action. Environmental relationships are constrained by the transformations effected through social practice; hence the characteristics of the time operating in these transformations.

Science as a co-operative perceiving

Authentic knowledge of such constraints is the result of objectivized social practice. Therefore, we re-vindicate science as a method of comprehensively accumulating reflected experience so that anyone may use it when it is needed. The most important question is 'not to understand the laws of the objective world so as to be able to explain it, but to use the knowledge gained in the elaboration of these laws to actually transform the world'.[18]

Work as an act of 'doing the world'

The real act of development comes 'when the body realizes that it can see'[19]: when we 'think with our eyes', as the Huicholes say. This means overriding the antagonism between production and reproduction with a dialectical relationship between manual (simultaneous) and intellectual (sequential) work. Manual work, as an instrument of sensorial liberation, is a means of disconnecting our cortex and getting in tune with the total environment to produce what is needed in a concrete situation.[20] Intellectual work is the act of coding such 'doing' with information as organized by the cortex. Work is therefore the action of using our means of production, communication and decision-making to transform the world, of which we are an integral part.

The logical mode of perception

THE BOURGEOIS REVOLUTION

To understand urban phenomena today, we must go back to the revolutionary epoch of the bourgeoisie, the English and French of the eighteenth century, with their new modes of production and discoveries of the structure of society, culminating in the Industrial Revolution in England and in the French Revolution of 1789.

The discovery of America prepared the international market, accelerating sharply the development of transport and industry. The bourgeoisie itself developed in this space and time.

This had three immediate consequences: (a) it brought into importance the control of distribution in the environment, valuing the trading activity above the others, and focusing intelligence in achieving this control; (b) it widened the horizon of society to the lowest manual workers; (c) it put production out of the context of consumption (industrial production was exported from, and agricultural production imported to, Europe) and overvalued consumption by linking it with the decision-making process. To each stage of the evolution of the bourgeoisie corresponded a political development: formation of a new State under the domination of feudal lords; armed association in the autonomous community; urban independent republic. Finally, the bourgeoisie 'conquered the exclusive hegemony of political power in the modern State'.[21]

THE STATE AS THE REALITY OF THE MORAL IDEA

The conceptual framework for this conquest of the bourgeoisie is provided by Hegel. His concept of 'civil society' implies the following power structure: the government of a modern State is a junta that administers the communal business of the bourgeosie through a system of abstract values (democracy, freedom, equality, etc.). These constitute the source of State authority over the citizens. The State is then 'the reality of reason'. This is the coercive element that provides the necessary order for environmental control (production and distribution sectors)—the source of the economic power of the bourgeoisie.

In this way the revolutionary bourgeoisie developed new values in terms of the quantitative relationships of objects to man, in that new space it had created called the international market.

In the feudal system we can conceive states of relative stability: the feudal lords providing protection to the peasantry, who in turn provide food for their lords. But the bourgeois system is in a constant state of conflict: society is supposed to develop by the antagonistic interaction between those who have capital and those who sell their labour force; to develop one element means necessarily to underdevelop the other. So, we optimize 'freedom', but not this freedom, here and now, but rather that freedom which is somewhere else, always.

THE CULT OF INTELLIGENCE

'You want to remain the same at the cost of your well-being.'[22] Going back to our example of man discovering fire, the bourgeoisie becomes aware—at the time of its revolution—of the relationships of production and reproduction brought about by the discovery of new land and industry. It then discovers that as long as it can keep control of the means of production, it can control market values, and that the experience gained in controlling the environment can operate as a coercive instrument: as long as the intelligentsia can partly hide the mental process achieved, it can disguise with abstractions the internal contradictions of bourgeois society. The bourgeoisie has, therefore, to develop such an intelligentsia. (Probably, for the pioneers of the bourgeois revolution, such a process was clear; afterwards they became unconscious of the perceptions that generated the bourgeois value system.)

Bourgeois society institutionalized a mental process strongly linked with cortical perceiving. Its system of values imposed on man the sequential time of the bourgeois endeavour: the isolation of parts of a process occurring simultaneously in reality. We could then perceive raw materials as products of an invisible labour, and relate such perception to the decision-making process.

The value system generated in this way departs from the belief that our acts are ordered from the cortex. Intelligence is dissociated from the biomental process social practice brings about. Environment and language have lost their unifying quality, as concrete reality is perceived and objectivized in terms of mystified abstractions; an intelligence aimed at systematizing such abstractions—rather than direct experience—has become the 'secret fraternity of political aristocracies'.[23] Such systematization makes possible the use of different sets of opposites in different contexts, making out of secrecy 'the definitional operative of security'.[24]

THE BOURGEOIS CITY CONSOLIDATES CORTICAL PERCEPTION

For our purpose, the most important consequences of the predominance of this mode in the organization of the environment are:

The increasing separation of the areas of production from those of reproduction. This generates two increasingly different modes of life: one by which matter is transformed, and the other in which man remains alienated from the transformation process.

The isolation and mystification of the areas where the decision-making process, affecting such transformations, is performed.

TIME IN THE PROCESS OF PRODUCTION

One kind of time is employed to produce something. Another, very different thing is the consumer's awareness of such sequential time. Yet another issue would be to question the value given to such time. The basic assumption is: the shorter the better; labour is an evil, hide it as much as possible.

The logical mode of perception has brought us to a lack of awareness of the labour process: when we buy an object we do not think about the time employed in making it, the elements it contains, where they come from, the degree of decisions (and the quality of them) taken by the worker in the production of it, etc. The relations of man both to matter and to other men that made the object possible by actively transforming the world remain concealed to the consumer. Objects have acquired an identity of their own, and slowly taken possession of man's space; we are out of touch with our environment, and the power which objects exercise over us is terrifying.

The disaggregating time of sequential planning

ENVIRONMENTAL SCIENCES METHODOLOGY

During the nineteenth and twentieth centuries the concept of 'sociology'—originated with Comte and propagated by Mill and Spencer—developed 'as a reaction against modern socialism';[25] that is, as a follow-up of the idealism-materialism antagonism.

The realm of pioneering scientific experience — the workshop context of the early English — was reconditioned by the bourgeois value-system. Gradually, it acquired a different context: the laboratories.

The isolation of science from social practice has reduced sensory stimuli for intellectual activity and severely damaged the operativeness of our senso-brain with all its implications.

The hegemony of a sequential mode has determined a methodology to tackle environmental phenomena.

The aim of research is to explain phenomena rather than to act upon them, transforming reality. This transformation is always going to be done by somebody else; we do not know *who*, and very often we do not care *how*, in concrete terms; what researchers are to do is to take the decisions about what should be done.

Its isolation from social practice implies the assumption of potential variables as fixed parameters; priorities are based on 'supposed needs' which are pre-established externally to the actual context of environmental action. Such priority parameters are then treated as a set related in terms of a one-to-one, cause-and-effect, relationship; only quantitative controllable changes are allocated for the future. In this manner, goals are defined and decisions for future social action are taken. Initial assumptions remain implicit throughout the whole process.

This methodology enables scientists to remain outside the phenomena and detached from their implications; it has mystified their activity (placing it above the probabilistic nature of matter in selecting hypotheses, above the viability of error that any of our brain-processing tasks has without the other), and alienated them from their own everyday life. We are reaching a state of saturation as we realize it is not science that we are practising, but the formalistic modification of an ill-defined kind of technology. Anyone who has been involved in the 'policy sciences' knows that much of this 'technical expertise' is nothing but jargon-ridden camouflage for political assumptions, or a pretentious disguise for simple rules-of-thumb.[26]

CHANDIGARH: THE DIALECTICS OF ENVIRONMENTAL WAR

The most important planning model emerging from this approach is that of the comprehensive development plan. Here, every activity of man is planned for, basing our assumptions on existing trends projected in sequential time, i.e. time is explicitly considered.

The application of the methodology mentioned above through this planning model has been disastrous.

Let us take for instance, the city of Chandigarh in Punjab, India. It was designed by Le Corbusier[27] in 1951, based on a set of abstract assumptions believed by him to be the key to the inhabitants' well-being. These he considered to be fundamental and unquestionable principles of urbanism applicable in any part of the globe and dealt with.

Controlling the basic function of the city, whether cultural, administrative, trading or industrial, not allowing a mixing of the two.

Within the city, separating the four functions of living, working, recreation, and circulation; and physically fixing their locations in the master plan.

Imposing formal visual controls on the containers (buildings) for the above activities to enable visual differentiations between them.

Linking the thus separated functions by a network of seven types of roads, again of different widths and specified uses, to allow fast movement of motorized traffic. Segregation of pedestrians from vehicular traffic formed an important aspect of the circulation network, and necessitated a second set of linear green spaces overlapping the road grid.

The Chandigarh plan was prepared within a week of Le Corbusier's first visit to India and approved by the Indian authorities soon afterwards. The main axes of the city are two wide boulevards based on the scale of those in Paris. We might ask if Le Corbusier was conscious at that time that the wide French boulevards were purposely built to allow for military control. In the same way, he completely ignored the historical development and complexity of Indian social formation. Marx had termed the original communal Indian village system the Asian mode of production. With the arrival of the British, the forces unleashed by British free trade dissolved and disorganized these small self-supporting organisms by exploding their economic basis, thus initiating a still ongoing process of social transformation.

Le Corbusier's assumptions had nothing to do with the historical reality reflected in people's environmental attitudes; he neither appeared to be conscious of it, nor did he propose a methodology of action to transform parameters of people's value-system into variables.

His objectives were unidimensional: to solve a particular problem as he saw it, and not to overcome a set of contradictions aggravated by colonialism. Le Corbusier actually believed that his plan would override economic differences; that the plan 'would take care equally of the peon and the minister' by locating them near each other. Hence his idealism and paternalism.

But in fact, the plan had no viable economic basis.

It relies on an artificial increase of land values to finance its physical development.

It allows for the provision of basic services through permanent institutions only; this demands high initial investment, thereby denying employment to those who need it most, and also forces high overhead costs on all, irrespective of income.

It segregates working and living activities, thus drastically altering the role of transformation of matter—production—within the Indian dwelling. This was done where the family was not a unit of consumption, but of production-reproduction interaction. Thus the house became a unit of consumption served by non-existent traders in unprofitable shops.

Today, twenty-five years later, the work of Madhu Sarin makes an important contribution to the evaluation of the master plan:[28]

The lack of economic viability of planned provision has generated a great variety of popular services which require minimum initial investment, provide employment to large numbers and goods that a majority of the population of the city can actually pay for. Hence the symbiotic relationship between the officially recognized and 'unauthorized' activities.

People providing these popular services build their own mud and thatch huts on the outskirts of the city. A special police force is responsible for the demolition of these huts, which the workers continuously rebuild. This dialectic interaction between the workers and the repressive State forces has stretched workers' ingenuity to the limit: if the police demolish houses, people put up tents which they quickly undo when the police appear; if stalls are prohibited, people put them on wheels, etc.

The planned dwelling is most unsuitable for the complementary services (additional sources of income or food) required as a safety valve in times of scarcity. It conceals, with a pseudo-environmental alternative, the lack of a viable economic alternative.

By ignoring people's biomental processes, and the organic time required to transform internal contradictions mapped on to the environment, Le Corbusier unsuccessfully tried to 'stop the other's world'. Chandigarh's master plan is an example of a static and abstract unscientific intervention; but also an encouraging example of people's resistance to alienation.

THE HOUSE AS A UNIT OF PRODUCTION

Among the most serious environmental problems we have in the world today is that of the habitat. We see it as an increasing problem, taking different expressions in countries with very different political and economic systems. And we wonder what the common denominators are among such countries that relate to the 'housing problem' common factor. In fact, the problem is defined as the need to build 'a house for each family'. One very important implicit parameter which is apparently forgotten is that this 'house for each family' is assumed as a unit for consumption. This character of it is never questioned. The author's position is that the housing problem will never be solved so long as the house remains a unit for consumption of hardware and software (including the information-processing consumed through television). But, when was a house a unit of production? In many parts of the world, it is today; nevertheless, this parameter is perceived as an element of a decaying culture. In the Western world's hegemonical consciousness, the breaking-up point of the old communal society occurred about 6,000–8,000 years ago with the introduction of stock-raising and agriculture.[29] Before the agricultural revolution, the households that women managed 'were not simply the first collective kitchens and sewing rooms, but the first factories, scientific laboratories, medical centres, schools and social centres'.[30]

The housing problem is not merely quantitative but also qualitative; it is not solved with the production—more or less efficient—of its hardware; its internal dialectic implies a qualitatively different relationship between the dwellers and their environment in terms of use value and not of exchange value. The solution implies the gradual qualitative transformation of the dwelling space so that consumption is no longer predominant, in order to recover, inside it, a place for social practice at the level of production, communication and decision-making. This effectively means to change the social dimension of the family.

This statement has two consequences for our study: (a) the house as a unit of production-reproduction implies that the man-woman relationship occurs within a context of transformation and not of conservation. This prevents us from alienating the rational and the intuitive modes of perception; (b) the only house that responds to our needs, and that we are able to maintain, is a house that we are able to transform.

In the same way that the bourgeois revolution widened the horizon of society to the poor, our society is widening its horizon to women and youth. This requires a qualitative change in the set of abstractions now used as reference parameters for our value system. Women are fighting to question the power structure in everyday life, to establish, ultimately, a different order: one where sensory information stimulates different biomental processes, bringing to the surface the faculty which enables man to perceive all opposites as alive, moving one into the other, transforming ourselves.

Viet Nam: The will to transform

It would have been very difficult to imagine, 'before Viet Nam', that it was possible to defeat such a highly developed technology as that of the

United States army with the means at the disposal of the Vietnamese people.[31] We are always amazed by the constant flow of energy among them, by their creativity, by their unity and ability to integrate everybody into a single global strategy, by their capacity to achieve concentration in a totally unstable situation.[32] This perceptual ability was gained through long training.

We saw similar phenomena in Mozambique and Guinea-Bissau during their struggle against Portuguese colonialism. The environment acquires the character of 'occupied cities'. Two armies of entirely different structural characteristics intervene in their transformation; one has a hierarchical power structure based on the pursuit of ideals (democracy, freedom, etc.) in strong contradiction with its environmental intervention. The other army changes targets according to a dynamic concept of working hypotheses determining an ever-changing set of priorities; it bases its strategy on the dynamics of the concrete situation as developed by people's socio-environmental intervention. Each one has, therefore, different territorial interests: one is fighting to preserve it, the other to transform it.

The destruction of buildings and social institutions goes with the construction of a defence apparatus and with the creation of new organisms to solve social problems (health, education) and to continue producing means of survival. Contrary to Warsaw—where people rebuilt the city after the war—this transformation is not temporary; it is gradual but irreversible; a new mode of perception is being created and new modes of production alter social relationships; the dichotomies mapped in the environment are overruled by the presence of a homogeneous social force.

But what is important is not getting used to bombing, or merely a new mode of perception. It is the process of everybody learning to systematize experience acquired through a simultaneous mode of perception. To the degree that everyone participates in this transformation, people create values out of objectivized social practice rather than defend a set of values, based on universal abstractions whose origins they are unaware of.

This consciousness of the social relationships that are generated by such processes enables us to maintain such values as dynamic norms that we use rather than dogmatic rules that we defend fearfully. This is what makes socio-environmental transformation irreversible.

We see such environmental phenomena spreading to increasingly larger world zones, though taking different expressions, not always implying a 'declared state of war'. Similar phenomena occurred in proletarian neighbourhoods in Chile (1970–73) and in Portugal (1974–75). By questioning power structures in everyday life, a great creativity is awakened to develop new techniques, and new social

organizations to satisfy their rapidly changing needs[33] long-term
theoretical studies are useless because qualitative changes in people's
consciousness constantly alter the conjunctional limits and transform
assumptions.

In our point of view, these phenomena aim at recovering man's
control over his own environment, control lost to the triumphant
bourgeois revolution, overriding the antagonism generated between
producers' and consumers' ways of being: 'to participate in production is
not only a duty, but a right for all and for everyone'.[34]

Conclusions

We all have the right and duty to participate in the transformation of
nature and of society. The instruments needed to co-ordinate such socio-
environmental action are most important for mankind's survival. The
value-making process cannot be determined in sequential time; it cannot
be imposed on, or sold, to people outside the sphere of the concrete reality
which generated such values. New values cannot be communicated
verbally; they can only emerge out of objectivized social practice.

All these imply new instruments for the intervention of environmental scientists: an awareness that wherever we are, we are
operating inside a global strategy for development[35] which has its own
social dynamic and timing; and that it is urgent to re-define the role of,
and the strategy for, the intervention of the intelligentsia in socio-
environmental phenomena. The very concept of a planning theory, which
defines the norms for social action, is severely questioned, precisely
because of the temporal experience it implies.

The objective of science is not to predetermine what we should
assume and what we should be searching for when faced with a particular
problem. In the first place, science is a means to detect real needs by
testing hypotheses in social practice; secondly, a means of collectively
deciding—according to the ever-changing priority problem—who is in a
position to solve it; finally, a means to establish a 'strategy of doing'.

All this implies a different concept of the time of environmental
action, which is fundamentally the time of social dynamics; and an ability
for pattern recognition rather than for forecasting. That is, a multidimensional intervention of environmental scientists at different levels of
qualitatively different temporal experiences. Our ability to intervene in
sequential processes (building, etc.) depends on the awareness of their
dialectic relationship with social processes in simultaneity.

What is required is a servo-ballistic intervention, using hypotheses of
work to enable us to change course according to inputs as we move along.
The role of the scientists in this process is to provide everyone with

instruments for systematizing experience, and to acquire from the base, instruments for simultaneous decoding.

We think it urgent to find tactical means to put our senso-brain into operation, to re-evaluate it, to understand its dialectics and to appreciate its value-making function.

A new dimension of temporal experience is undoubtedly a means for coming to terms with the required transformation in our mode of perception. This might give us the humility and the tools we urgently need to overcome the profound antagonism between man and his own nature in today's world.

The author acknowledges the important contributions to this chapter made by Andreas Adam, Madhu Sarin, Chris Crickmay and many others.

NOTES

1. Robert E. Ornstein, *The Psychology of Consciousness*, San Francisco, Calif., Freeman, 1972.
2. R. L. Gregory, *Eye and Brain*, 2nd ed., London, World University Library, 1973.
3. Our eyes are primarily detectors of movement which is coded into the neural activity of the retina.
4. The biologically oldest part of the brain which processes sensorial information (striate).
5. If we balance a wooden dowel on the index finger of each hand while we speak, the balancing time of the left hand increases with verbalization; for the left part of the body—not directly linked with the right hemisphere cortex—verbalization is perceived as sound which stimulates its performance.
6. Andreas Adam, 'Konfrontationen' (contribution to an exhibition), Zurich, Department of Architecture, ETH, 1975; also, *Archithese*, No. 17, 1976; and John Berger, *Ways of Seeing*.
7. To say that language is based upon linearity is to restrict verbal communication as exclusively related to a memory of non-visual characteristics. This depends on the logic of the language structure. Poetry has a language structure bridging to a pre-logical mode.
8. From experiments on perceptual isolation we learn that the absence of sensory inputs can very badly damage our brain; our experience of duration shortens. With drugs (peyote, marijuana) we can experience the opposite process, disconnecting our cortex. The intellect cannot check the validity of intuition as this has its own logic and validity depending on what we are using the whole process for.
9. Freud's symptoms of repression and denial, in which verbal mechanism has no access to emotional information.
10. Karl Korsch, *Three Essays on Marxism*, London, Pluto Press, 1971.
11. Karl Marx, *Textos Filosoficos*, p. 110, Lisbon, Biblioteca do Socialismo Científico, Estampa, 1975.
12. Carlos Castaneda, *Journey to Ixtlan*, London, Bodley Head, 1973.
13. C. Lévi-Strauss, *Tristes tropiques*, P. 188–90, Paris, Union Générale Edition, 1955 (10/18 series).
14. There is a popular term, among Mexican people, for this instrument of domination: 'white violence'.

15. Writings of the time of the Spanish conquest—like those of Bernal Diaz del Castillo—described how words quite often seemed to hurt Indians much more than physical blows.

16. Carlos Fuentes, *Tiempo Mexicano*, p. 37, Mexico City, 1972, Cuadernos de Joaquin Mortiz.

17. A popular Chilean term to qualify reactionary attitudes was *momio* (mummy); it obviously relates to the lack of energy and willingness to transform themselves among reactionary groups.

18. Mao Tse-tung, *Obras*, III, 'O Proletario Vermelho', Lisbon, 1975; *Sobre a Practica*, 1937.

19. Castaneda, op. cit.

20. E. F. Schumacher, 'Buddhist Economics', *Manas*, Calif., Vol. XXII, No. 33, 1969.

21. K. Marx and F. Engels, 'Burgueses y Proletarios', *Manifiesto del Partido Comunista*, p. 32–3, Moscow, Progreso.

22. Castaneda, op. cit.

23. V. Marchetti and J. D. Marks, *The CIA and the Cult of Intelligence*, London, Jonathan Cape, 1974.

24. 'For the cult of intelligence fostering stability may in one country mean reluctant and passive acquiescence to evolutionary change, in another country, the active maintenance of the status quo; in yet another, a determined effort to reverse popular trends towards independence and democracy.' Marchetti and Marks, op. cit.

25. Korsch, op. cit.

26. British Society for Social Responsibility in Science, London; see pamphlets such as 'The Technology of Repression', 1974.

27. Le Corbusier, *Oeuvres Complètes, 1952–57*, 3rd. ed., Zurich, Boesiger, Girsberger, 1957.

28. Madhu Sarin, *The Chandigarh Experience, 1951–1975*, London, Development Planning Unit (Bartlett School of Architectural Planning), School of Environmental Studies, University College London, 1975. Her work is based on her own social practice in Chandigarh; it shows very clearly the dialectic interaction between the planned and the unplanned parts of the city now in existence; and the continuous interaction of people with the transformation of matter in an environmental-responsive manner, as well as explaining the economic basis for such attitudes.

29. Sabina Roberts, 'Revolutionary Dynamics of Women's Liberation' (pamphlet), London, League for Socialist Action, 1974.

30. R. Briffault, *The Mothers*, 1927. See also the work of V. Gordon Childe, Sir James Frazer and Otis Tufton Mason.

31. Working group, *Vietnam—Ni Todas las Bombas del Mundo*, Chile, Quimantu, Documentos Especiales, 1973.

32. Ho Chi Minh, *Selected Articles and Speeches, 1920–67*, London, Lawrence and Wishart, Ltd, 1969.

33. In Chile, the United States blockade regarding spare parts for industrial machinery obliged workers to create other ways to accomplish part of the assembly-line process, or even to make their own spares by hand.

34. Samora M. Machel, *Estabelecer o Poder Popular para Servir las Massas*, Lisbon, Frelimo, 1974.

35. Henri Lefebvre, *La Revolution Urbaine*, Paris, Gallimard, 1970.

TIME IN RURAL SOCIETIES AND RAPIDLY CHANGING SOCIETIES

F. N'Sougan Agblemagnon

The problem of time in Africa

In this volume it is entirely appropriate that provision should be made for a contribution treating the subject of time and the sciences from the standpoint of socio-cultural studies.

Science may be the paramount hope of mankind, but the actual control of science remains a challenge that man has not yet been able to meet.

In other words, however much science impresses us with its wondrous works, the sociologist and the anthropologist—and indeed the philosopher—must, more than ever, study the ways in which it makes an impact on society.

The socio-cultural study of time in rural societies and rapidly changing societies offers us a vivid example of the crisis at present affecting much of mankind, with regard both to concepts and to social models.

Accordingly, when broaching the subject of time from the standpoint of rapidly changing societies and from that of rural societies we should, in fact, given sufficient time and space, deal with all aspects of the problems raised by this concept and bring out the specific implications of all sociological time systems.

We must, however, make a few reservations regarding the scope of the subject.

First, rural societies are diverse. Viewed more closely, the rural societies of the developed countries do not always resemble those of what are termed the developing countries. And the numerous symposia of CENECA[1] I have attended have taught me to perceive the diversity of rural societies in the present-day world. The same applies to what are known as the developing societies and to all structures that we may regard as undergoing rapid change. Therefore these notions of 'rural societies'

163

and of 'rapidly changing societies' are in this context merely a convenient way of taking in a whole range of situations that are in reality infinitely varied and differentiated.

Even though one can none the less attempt to adumbrate the impact of the various time systems on the life and culture of men and communities in rural societies, on the one hand, and in rapidly changing societies, on the other, for our present purposes we shall not draw any distinction between these two types of society.

In many cases, and Africa is no exception, it is the same reality which undergoes the impact of such change. Our cities still bear the stamp of the countryside and our urban leaders have a considerable hand in village affairs. There is no clear dividing line between city structures proper and traditional village structures. There are many areas of encounter or, in any case, instances of permanent though often antagonistic existence side by side. We can, however, define in broad outline what rural African societies represent today in relation to time.

I have, in previous studies, clearly shown by means of *ad hoc* analyses how these rural African societies are today singularly characterized by encounter and dualism, with regard both to models and to concepts.[2]

In other words, the same will be true of time. We will find specifically traditional aspects, stemming from traditional cultures which are still very much alive and, on the other hand, areas where modern time has crept in, be it with reference to the time of the planner or to the notions of time conveyed by the particular sciences thereby introduced into the social setting, including the rural one, through development plans.

As to urban reality, our cities may closely resemble others elsewhere in the world; and we may sometimes be misled into believing that extravagant or superficial displays of grandeur and modernity are anything other than show, and thus be tempted to draw over-hasty analogies with the developed industrial societies.

But African cities cannot yet be viewed from the same angle as cities in Western Europe or North America, even though concrete may be the prevalent factor and administrative power may tend to supplant traditional power. The African city, in terms of organization, structure and daily life, partakes deeply of traditional life, and there are relations of periodical or permanent symbiosis between it and the environment, and between it and the traditional villages. This means that African man, even as a city-dweller, cannot escape the ties of the traditional milieu.

As has been shown in studies carried out on university élites of the city of Ibadan by Professor Peter Lloyd,[3] there is frequent contact between town-dwellers and villagers, if only to maintain community ties with relatives who have remained in the village.

But the town, perhaps to a greater extent than the village, offers, so far as time is concerned, a truer and more dramatic picture of the tensions that arise from these twofold conceptions of time and from these clashes between patterns of thought which reflect different epochs and are sometimes directed towards opposing ends.

This study will therefore chiefly set out to explain, with ample examples, the traditional African notion of time and how both African communities and African individuals, in villages and in towns, are now living these new tensions arising from the silent, or sometimes violent, confrontation of very different time systems.

The problem of time in Africa is certainly a fundamental one. What is more, it is directly linked to that of the environment and that of space.

All the main myths of African societies give what I would call a synthetic and unitary definition, taking in both time and space together. Primordial time is often, and at the same time, a primordial space. And the symbols, particularly that of the couple, which we often find bound up with it, imply that the space-time duality is lived by Africans, at whatever level they may be situated.

But today the new factor, the new parameter, is the introduction of breaks, the introduction of conflicts, and the introduction of models or patterns of thought that undermine both the social system as a whole and the basic concepts which went to make up the harmonious and harmonizing core and fabric of traditional societies.

In other words, Africa is today experiencing, in both village and city alike, an overlapping of times, and its entire psychological impetus is directed towards the frantic quest for ways of bringing its various notions of time back into harmony.

African man, like African society of today, is subjected to conflicting forces and even frequently torn asunder. Then the very models which still seem intact are beginning to yield a little within, which means that they must ultimately be overhauled or find the inner strength to make themselves whole and entire once more.

African time is not yet a time of disarray but it is, in several cases already, a time of rupture, of scission. And this is evidenced at all levels when we consider the major phases of the various time systems.

First of all, astrophysical time has always existed in African societies and there is no need for a community to build up a system of physics or astronomy, in the modern sense of the term, to have an astrophysical time. Incidentally, we somewhat underrate such societies if we disregard certain scientific aspects of their astrophysical time.

Measurement, in the strict sense of the word, is not always lacking in this respect. The heavenly bodies do of course serve as celestial timekeepers, as it were. Both their system and their configuration possibly provide a foretaste of the future, while their supposed influence may be held to afford a basis for rules of divination. In any case, the traditional calendars in their relative accuracy show that the heavenly bodies have even served to punctuate real physical time and remain sound benchmarks for man, even in traditional societies.

But the main importance of this calendar of the heavenly bodies seems to be that it furnishes a ground for the first major concepts which were to be used later by all philosophies and all science. I am referring to the theory of the world as a cosmos and as a complete entity, and the way in which myths—which to a certain extent are comparable to operational concepts—really act in the manner of concepts when it comes to formulating a world view and ordering what is deemed to be the reality of the universe.

Mythical reality, far from being an isolated line of thought, reveals or enfolds elements which are already pre-scientific and give man a certain hold over the perceived world.

At this level, the time of the soothsayer certainly takes precedence over the time systems of the scientist. But does this mean that this reality only truly belongs to the world of what are called the developing countries, of societies simply termed traditional? Or, on the contrary, should not the world of the myth be regarded as a world just as permanent and old as man's attempt to rationalize his knowledge and to organize scientific know-how and power? The fact that what I would call a mythical order persists in what are at present the most advanced and most industrialized societies is a clear indication that man's thought wavers in fact between science and dreams.

And we shall see that the contemporary African world, be it in the village or even in the cities, partakes in this duality comprising on the one hand science, namely, an effect of rational organization, and, on the other, dreams, namely, a future projection which goes beyond present scientific certainties.

In traditional African societies, biological time has also played a major part. We may almost speak of this biological timekeeper which was, to be sure, man's first tutor in that it reveals to us rhythms and, through such rhythms, the divisions of time, or in any case of a time which can be split up into units. Are not biological pulsations themselves the first step to an understanding of music? We can in this respect regard our hearts as veritable metronomes. Are we Africans not introduced to the pace of time as infants slung on our mother's back, listening to her muffled heart beat?

Africa has turned biological time to account and made use of its in all patterns of behaviour, the most modern no less than the oldest and deepest-rooted. Our introduction to science, our introduction to knowledge and even social power are, to a certain extent, bound up with this biological aspect of the world.

Psychological factors

In African societies age is still a factor to which great weight is attached. The relationships deriving from birthright have become firmly established, both as between generations and as between individuals of the same generation. It is possible that we have here a concept or a factor which had a noteworthy influence on social and political conceptions, or quite simply ideas of authority in African societies, whatever the types considered. It is therefore impossible, in our present framework, to present all the implications of biological time in Africa.

But I also think that as regards psychology, the way in which time manifests itself to the African is reflected in a more interesting and clearer way in terms of the relationship between the social and the collective. Although mythical time certainly loomed large in traditional time, still more noteworthy was the emphasis placed on the social and the collective, to the detriment of the individual.

In other words, though the time of individual consciousness is an actual reality, it is always experienced in a social and collective context, so that the individual perceives himself more as a social, collective being than as an individual being in the sense of a psychology of introspection. In other words, the primordial conception of time is such that the African, even when he has evolved in new cultures, remains attached to the implications of this traditional time, and is still distinctly marked by the collective and social dimension of time.

Sociology and anthropology offer us, in actual fact, a better perception of African time in that we see the cycles, the periods and the various time systems of the different social groups emerge more clearly. And in this process, it is not just differences between time systems that appear, but also discontinuities, opposing trends and contradictions.

The African world is therefore not a unidimensional world, and still less an immobile world, as naive anthropologists might have wished, but on the contrary, a world under stress, a world subjected to conflicting forces, and a permanently changing world.

Research, in that it brings us data from different epochs, may all too easily result in widening the gulf between these different time systems. At the same time, it is becoming less easy in African society for the African man of today to adjust himself to time and to culture since we are no longer very sure which culture we are in, where traditional culture ends and where modern culture begins.

These breaks, these disharmonies and these new tensions are an essential dimension of time as now experienced, and futurology itself merely gives expression to contradictions that have become apparent.

Between the divinely inspired futurology of the priest or soothsayer of traditional society, and the rationalistic, statistical futurology of the modern economist, there is a gulf. The problem of time in rapidly changing African societies, and traditional rural African societies undergoing modernization thus offer us an ideal opportunity of grasping, dramatically, the problems arising from the duality of time systems. This means 'studying this problem of the African consciousness of time, not vaguely and academically but in a practical manner, that is to say, on the basis of precise questions.

First of all, what today is the situation, or rather the relationship, of tradition and change?

Sociologists and anthropologists can be said to hinge their entire approach on these two concepts. However, it seems to me that a very critical attitude is called for in regard to them.

Tradition is not, as is sometimes thought, a dead, congealed and unvarying reality. Previous studies, particularly on the material of folk tales in traditional African societies, show that tradition itself has always evolved and is still subject to continuing development. Although change impinges more readily on the attention than the stable core of a society or of an institution, it remains none the less true, to use a formulation I have already adopted on an earlier occasion, that 'the unchanging elements in any society are just as significant as those that change'.

In other words, we shall resist the temptation here of seeing in change, and change alone, the most significant aspect of the present trend of African society.

As to change, we know that it often operates without proper guide-rails. We do not always impart to it the appropriate meaning. In other words, unlike the ultimate goals of traditional time, the ultimate purpose of the changes currently affecting African societies undergoing transition, often escapes us, or sometimes remains unknown to us or, in any case, undeciphered.

As to the speed of change and problems of adaptation, we cannot over-emphasize the fact that many changing societies and many rural societies in the Third World, particularly in Africa, can no longer draw on what I have frequently called time for reflection, in order to make choices that are sound if not *de facto* at least *de jure*.

It is therefore to be regretted that the speed of change often, to a certain extent, obliterates the latent significance of the result of change and, in many cases, deprives man, as actor and agent, of a sense of its meaning, and thus divests him of the power to control change and, with it, the time against which it takes place.

This is a serious problem which naturally prompts the question: what degree of immobility, what admixture of tradition and of change, must be imparted to, and accepted in, social development in order to render it controllable and in keeping with the ultimate purposes assigned to it?

We must—and these changing rural societies in Africa give us many an illustration of this—above all, control innovatory models; otherwise there is a danger of their acquiring a dynamism of their own, quite unrelated to the true objectives set as ideals for society.

Social structures

The problem of social structures in rapidly changing rural African societies remains all-important.

We know already how complex the situation in regard to structures in a present-day society may be, so that a study which set out to analyse these structures in a rigorously scientific manner would reveal extremely varied degrees, levels and ages. Even when traditional models and new models are mutually receptive, it is not possible to judge in advance the quality of the results obtained; this is all the more so when a society is like a vast experimental plot into which we introduce innovations almost daily and, sometimes without adequate and appropriate checks, import foreign models. In other words, even the most pertinent sociological analysis cannot reveal the true dimension of the dualities and conflicts engendered by the models operative within social structures.

Of course, new values come into being both in daily life and at work. But here, too, there is a juxtaposition of different time systems, there is a juxtaposition of models from different epochs, and there is sometimes a juxtaposition of behaviour patterns in one and the same person.

Will not the illiterate worker try, in the factory, to imitate his Western colleague but, back at home, resume in full his traditional behaviour in his relations with wife, children and neighbours? This life on two planes at once, if not on several planes at once, is a common phenomenon to be observed in these rapidly changing rural African societies. Sometimes attitudes, but also moral and aesthetic values, are unavoidably modified or traumatized as a result. Here it is not the best models that gain the upper hand, and there is, regrettably, a certain moral deterioration among certain groups of young people in the cities.

In these circumstances, the environment is not that context which truly integrates, but a context raising problems for the individual and for the group.

Even though the African is not always, is not yet incorporated in a purely urban structure, the way in which certain innovatory economic structures have been introduced into the general environment already exerts a definite influence on this setting, and on human behaviour patterns.

I have shown, particularly in a paper presented to the second world congress convened in Brussels on 5–7 April 1976, under the title 'Temps

Libre et Accomplissement' for the purpose of drawing up a 'leisure charter', that the work-leisure relationship affords, in this connection, a very interesting clinical insight into the crisis which African concepts of time are currently facing. As I wrote on that occasion, 'The notion of "free time and accomplishment" is of paramount importance for African societies, be they traditional or modern'. And we are well aware that, nowadays, time is a time disrupted and fragmented, a far cry from the living time of traditional societies—the time of totality and integration; no longer is there the same harmonious alternation of time for work and time for rest. We are today far removed from that time of integrating harmony, at once ontological and sociological, which I described in my study on time in the Ewe culture.[4]

In traditional society, the time of rest was generally devoted to the divinity, while today it tends to be given over to more profane principles, to biological repose and not simply to contemplation and self-communion. And we know that the conception of rest was part of a far broader structure in which religious and collective elements played a greater part than the purely temporal, personal and sensual aspect of reality.

In the paper to which I have already referred, I went on to say that the purpose was not to 'take advantage of the world, and especially of the blessings of the earth, to use things, and time, for our enjoyment, but to become one with time'.[5]

There is, so it seems, a vast difference between the all-embracing time which enfolded the individual in all his aspects and made him part of a global reality, that is, ultimately related him to a cosmic time, and the fragmented, scattered, convulsive and conflictual times which we experience in the present-day transitional African societies.

As I have said, the intention of the traditional celebration in African societies was not personal enjoyment, but it was, in a sense, a way of reintegrating the individual with society and of activating or intensifying inter-personal and inter-community relations.

Traditional African time in fact integrated the individual, whereas modern time tends, on the contrary, to dissociate him from the social setting and isolate him.

It is accordingly very important to emphasize how actual examples, indicating the way in which time is lived by the various social categories and the various social circles in Africa, illustrate the difficulties of the problem of time in present-day African societies.

The example of the worker to which we have just referred is not merely a question of behaviour patterns; it is also relevant to the way in which the present is assessed and the future imagined.

Likewise, the official who follows the Gregorian calendar is, to a certain extent, also torn between the so neatly distributed time of his occupation from 1 January to 31 December of the administrative year, and the other, traditional cyclical or periodical calendars.

The achievement of a *modus vivendi* between these different time systems amounts, in some cases, to a real feat, both for individuals and for African society at large.

Similarly, young people provide ideal material for case-studies on the way in which they transpose time into actual experience. The case-study I conducted on young out-of-work literates at Tsevie in 1960 show-ed how their feeling of having no place in society was mainly responsible for their impression of diminished social status, and that what counted for them was, above all, to remain on good terms with other generations, particularly that of the parents and that of the children.

Conclusion

So time—this fragmented African time—is lived in a fragmented manner but also in fits and starts, according to the models that present themselves to the consciousness, and according to the actual situations encountered by, or involving, the individual. This permanently distorted and contingent character of the present-day African experience of time remains the most formidable obstacle to a unification and integration of personal experience.

The problem, then, is how to instil harmony into this conflictual time, how to convert this time of disharmony into a time of harmony and how, in working life and in day-to-day social relations, to bring about true fulfilment, how to transform this time of crisis, this time of aggravated dualism and this time of disharmony into a time of harmony and accomplishment. This seems to be the problem facing the student and the illiterate peasant alike.

The rural world is still more traumatized by the fact that the peasant does not always possess the keys, even the scriptural keys, to the system. He undergoes the models. He undergoes the explanations given to him. He cannot extend his own reflection. He cannot consult books and reformulate his problems, or systematically evolve a personal, critical and comprehensive view of the system.

Another still more striking illustration of the crisis with which African time is faced concerns the operations of planning agencies and modern technological design offices. Indeed all those, in whatever social category, whom we may legitimately regard as decision-makers or people in positions of responsibility, be they administrators, politicians, economists, teachers or analysts, are living this time of scission and contradiction. Village and town alike are very sensitive observation points in this respect: the town because it holds within it and magnifies all the psychological conflict experienced by the individual; the village because it contrasts, sometimes crassly, models of utterly opposed or contradictory origins, ages and basic purposes.

The changing structures themselves and, among them, even those which appear to be successful, do not always provide us with a key with which to assess thier validity in time. They are simply a product of the present. They do not really answer for the future, or for their own future.

Hence the fundamental, cardinal question to which we have still to seek an answer is that of the ultimate purpose of social time.

For there lies the major difference between African traditional time and modern time systems. There is an essential difference between this mythical time, tending in all circumstances to integrate rather than to isolate, and these chequered time systems which, on the contrary, divide and oppose. Hence the conflict between the time of the individual being and the time of existence as a whole is more than ever heightened. The time of contradiction far outweighs the time of accomplishment. The time of miscellany and multiplicity prevails over the time of entirety, of integration and of harmony, and inner divisions destroy inner peace.

In other words, this time is an uncontrolled time, a time which, far from satisfying the individual, isolates him and even assails him. It is a far cry from the traditional course of social and cosmic time, tending towards comprehensive and full integration, not only of the individual, but of society at large.

Nevertheless, do not these phenomena we observe, these fierce conflicts revealed to us by Africa's rural societies and rapidly changing societies, hold out some hope to us?

We can, in any case, see for ourselves a certain convergence of the time of science and the time of actual experience. Yes, the way in which science is becoming a consumer item is, in itself, a certain means of reconciling the time of science with the time of reality as in fact experienced.

We are witnessing a similar convergence of the time of modern futurology and of mythical time. Moon flights are of course a scientific reality, but the bulk of the inhabitants of the earth—in fact, everyone except the cosmonauts themselves—live the lunar scientific exploits only on an imaginary, and almost mythical, plane. This applies even more so to unmanned flights. Furthermore, do not the most exact sciences, beginning with physics and biology, portray in their most relevant descriptions, a world closer to fiction than to reality, in terms of real perception? And is there not finally a conjunction between scientific time and oneiric time, the time of the dream?

Science should be able to become a dream for mankind, a new dream. And science should not merely set out to clarify hidden realities: it should endeavour, on the physical plane as on the psychological and sociological plane, to prepare for us an environment and a time which incorporate us, a time which caters for the dreams of man and the dreams of society.

NOTES

1. Centre National des Expositions et Concours Agricoles, 19 boulevard Henri IV, 75004 Paris.
2. 'Conflits de Passage et Sociologie des Jeunes États Africains: Le Cas du Togo', in *Civilisations*, Vol. XVIII, 1968, No. 2, p. 232–46.
3. *The New Elites of Tropical Africa, Studies Presented and Discussed at the Sixth International African Seminar at the University of Ibadan, Nigeria, July 1964*; edited with an introduction by P. C. Lloyd; foreword by Daryll Forde, Oxford University Press, 1966.
4. 'Du Temps dans la Culture Ewé', contribution to the first Congress of Negro Writers and Artists, Paris, 1956.
5. 'Dynamique Sociale et Problèmes de l'Utilisation du Temps Libre en Afrique Aujourd'hui', second world congress for a 'Charte des Loisirs', on 'Temps Libre et Accomplissement', loc. cit., Brussels, 5–7 April 1976.

NOTES ON THE AUTHORS

AHMED M. ABOU-ZEID. Professor of Sociology and Anthropology, Head of the Anthropology Department, Dean, Faculty of Arts, University of Alexandria, Egypt.

JOSEFINA MENA ABRAHAM. Born 1941. Graduated in architecture, Mexico, 1965; postgraduate studies in urban sociology and social anthropology, Sorbonne, University of Paris, 1966; Bartlett School of Architecture, University College London, 1969. Tutor at School of Architecture, Kingston Polytechnic, Kingston-on-Thames, England, 1969–72. Professional practice in Mexico, France, England, Portugal, mainly with housing. Since 1975 has worked with Serviço Ambulatório de Apoio Local, Departamento do Fundo de Fomento de Habitação, in working-class neighbourhoods around Lisbon, Portugal, following similar work in South America.

Her publications include articles in *Oasis*, *Arquitectos de México*, *Science for People*, *Impact of Science on Society* (Unesco); also editorial assistant for *Design Magazine*, 1973.

F. N'SOUGAN AGBLEMAGNON. Born 1929, Ahépé, Togo. *Licencié ès Lettres*; *Diplômé d'Études Supérieures de Philosophie*; *Certifié d'Ethnologie*, Faculty of Science, Paris University; Ph.D. in sociology. Ambassador, Permanent Delegate of Togo to Unesco; *Chargé de Recherche*, Centre National de la Recherche Scientifique, France, since 1956; Director, African Laboratory of Joint Research and Interdisciplinary Studies; President (1975–78) and Vice-President (1979–80) of the Board, International Bureau of Education, Geneva; Vice-President of the Board, Centre for the Co-ordination of Social Science Research and Documentation for Africa South of the Sahara (CERDAS).

Published articles include: 'L'Afrique, la Métaphysique, l'Éthique ... ', *Comprendre*, Venice, 1961; 'La Vie Africaine', *Afrique Africaine*, Lausanne, La Guilde du Livre and Claire Fontaine, 1963; 'Mythe et Réalité de la Classe Sociale en Afrique Noire: le Cas du Togo', *Cahiers Internationaux de Sociologie*, Vol. XXXVIII, 1965, Paris, Presses Universitaires de France; 'Aspects Démographiques de la Croissance Économique: le Cas du Togo', *United Nations*

World Congress on Population, Belgrade, Yugoslavia, 1965, New York, United Nations; 'The African Child', in F. R. Wickert (ed.), *Readings in African Psychology*, African Studies and International Programs of Michigan State University, 1967; 'Problemas de la Independencia en Africa Negra', *Anuario de Geografía*, Mexico City, Universidad Nacional Autónoma de México, Facultad de Filosofía y Letras, 1966; 'Tradition et Mutation dans les Sociétés d'Afrique Noire', Neuchâtel, Seventh Symposium of the International Association of French-Speaking Sociologists, 1968; 'Sociologie Littéraire et Artistique de l'Afrique', *Diogenes*, No. 74, Paris, 1971; 'La Responsabilidad de la Mujer Africana como Esposa y como Madre', *Mujer y Entorno Social*, Madrid, Fundación General Mediterránea, Patronato José Ferrer, 1976; 'Musique et Société en Afrique Noire' and 'Danse et Société Africaine', *Monde de la Danse*, Vol. I, No. 1, 1976.

PAUL FRAISSE. Professor at the Université René Descartes (Paris V); Director, Laboratory of Experimental and Comparative Psychology, École Pratique des Hautes Études (associated with the Centre National de la Recherche Scientifique, France). Président, 3ème Section de l'École Pratique des Hautes Études. His studies are concerned mainly with the problems of perception and those of the psychology of time and rhythm.

Main publications include: *Manuel Pratique de Psychologie Expérimentale*, Paris, Presses Universitaires de France, 1966, 4th ed. 1974, translated into six languages; *Les Structures Rythmiques*, Paris, Erasme, 1956; *Psychologie du Temps*, Paris, Presses Universitaires de France, 1957, 2nd ed. 1967, translated into English and Japanese; *Psychologie du Rythme*, Paris, Presses Universitaires de France, 1974, translated into Spanish and Italian; co-editor with J. Piaget of *Traité de Psychologie Expérimentale*, 9 vols., Paris, Presses Universitaires de France, 1963–66, 3rd ed. 1975, translated into nine languages.

FRANK GREENAWAY. Born 1917. Educated at Jesus College, Oxford, and University College London. Keeper of Department of Chemistry, The Science Museum, London; (Honorary) Reader in the History of Science, The Royal Institution of Great Britain; Secretary of the International Union of the History and Philosophy of Science (Division of History of Science), 1972–77.

Publications include: *Science Museums in Developing Countries*, Unesco, 1962; *John Dalton and the Atom*, 1966; and official publications of the Science Museum. Has published critical editions of *Lavoisier's Opuscules Physiques et Chimiques* and *The Royal Institution Managers' Minutes*; has edited official publications of the Science Museum and made contributions to periodicals of museology and history of science. Has organized numerous international scientific meetings including the Unesco symposium, 'Time and the Sciences' (February 1976), at the Royal Institution, London, and edited its proceedings which form the subject of this volume.

MÜBECCEL B. KIRAY. Born 1923, Izmir, Turkey. Ph.D. in sociology, University of Ankara, 1946; Ph.D. in anthropology, Northwestern University, United States of America, 1950. Head, Department of Social Sciences, Middle East Technical University, 1965–72. Morris Grusberg Fellow, London School of Economics and Political Sciences, 1974; since 1975, she has been professor at the Istanbul Technical University.

Publications include: *Ereğli: Ağir Sanayiden Önce bir Sahil Kasabasi*, Ankara, DPT, 1964; *Yedi Yerleşme Yerinde Sosyal Yapi Araştirmasi*, Turizm Bakanliği, 1965; co-author with J. Hinderik, *Social Stratification as an Obstacle to Development*, New York, Praeger, 1970; editor of *Social Stratification in the Mediterranean Basin*, The Hague, Mouton, 1971; *Örgütleşemeyen Kent*, Ankara, Türk Sosyal Yayinlari Bilimler Derneği, 1972; and various articles in American, British, Dutch and Turkish publications and periodicals.

JOHN MCHALE. Born 1922 in Scotland, deceased 1978. Educated in the United Kingdom and the United States, he held a Ph.D. in sociology. Founding Director, Center for Integrative Studies, College of Social Sciences, University of Houston, Texas. As an artist and designer, he exhibited widely in Europe after 1950; his work included graphics, exhibit design and films. Fellow of the World Academy of Art and Science (Secretary-General and Vice-President); The Royal Society of Arts; The New York Academy of Sciences; and the American Geographical Society. He was awarded the Médaille d'Honneur en Vermeil in 1966 by the Société d'Encouragement au Progrès; the Knight Commander's Cross, Order of St Dennis in 1974; and an Honorary Doctorate of Law from Ripon College, Wisconsin, in 1977. He was a member of the American Sociological Association; Society for the Advancement of General Systems Theory; Colorado Archaeological Society; and the World Future Studies Federation (Vice-President). He served as consultant to several international agencies.

He published extensively on natural resources, on the impact of technology, on culture, mass communications and the future. His works include the following: *The Changing Information Environment*, London, Paul Elek (Scientific Books), 1976; *Westview Environmental Studies*, Vol. 4, Boulder, Colo., Westview Press, 1976; *World Facts and Trends*, New York, N.Y., Collier Books, Inc., 1972; *The Ecological Context*, New York, N.Y., George Braziller, Inc., 1970; *The Future of the Future*, New York, N.Y., George Braziller, Inc., 1969. In addition, he co-authored with his wife and colleague, Magda Cordell McHale: *World of Children*, special edition, *Population Bulletin*, Vol. 33, No. 6, 1979, Washington, D.C., Population Reference Bureau, 1979; *World's Children Data Sheet*, Washington, D.C., Population Reference Bureau, 1979; *Children in the World*, a chartbook for Population Reference Bureau, Washington, D.C., 1979; *Basic Human Needs: a Framework for Action*, New Brunswick, N.J., Transaction Books, 1977; and *Futures Directory*, Boulder, Colo., Westview Press, 1977.

A. NEELAMEGHAN. Born 1927. Professor and Head of the Documentation Research and Training Centre, the Indian Statistical Institute, Bangalore, India, from 1965 to the present. Currently Project Co-ordinator of the Unesco-United Nations Development Programme (UNDP) inter-country project on Postgraduate Training Course for Science Information Specialists in Southeast Asia, University of the Philippines, Quezon City. Majored in physics, Madras University, 1947; Postgraduate diploma in library science,

Madras University, 1950; M.A. in library science, George Peabody College for Teachers, Nashville, Tenn., 1954, with philosophy of science, audio-visual communication and vital statistics as allied studies. Post-M.A. specialization in medical bibliography and literature, Columbia University, New York, N.Y., 1954.

Librarian, Pasteur Institute, Coonoor, 1948–49; Madras Medical College, 1951–56; Library and Technical Information Centre, Hindustan Antibiotics Ltd., Pimpri, 1956–62; Reader (Associate Professor), Documentation Research and Training Centre, Bangalore, 1962–65. Visiting Professor, School of Library and Information Science, University of Western Ontario, London, Canada, 1973–74. Special lectures on documentation and information at the schools of information studies in the following institutions: University of Pittsburgh, Penna., Syracuse University, N.Y., University of Rhode Island, R.I., George Peabody College for Teachers, Tenn., University of Toronto, University of Ottawa, University of Minas Gerais in Belo Horizonte, University of Brasilia, the National Scientific and Technical Documentation Centre in Budapest, Consejo Nacional de Ciencia y Tecnología (CONACyT: National Council for Science and Technology) in Mexico City, etc. Chairman, Advisory Committee, General Information Programme, Unesco and United Nations Intergovernmental Programme for Co-operation in the Field of Scientific and Technological Information (UNISIST), 1974 to date; Chairman, Classification Research Committee of the International Federation for Documentation, 1973 to date; Member, Development Science Information System (DEVSIS) Study Team. Advisory and consultancy missions to, or on behalf of, Unesco, United Nations Industrial Development Organization (UNIDO); United Nations Asian Development Institute, Bangkok; United Nations Economic and Social Commission for Asia and the Pacific (ESCAP); International Development Research Centre, Ottawa; Consejo Nacional de Ciencia y Tecnología (CONACyT), Mexico; Industrial Development Centre for the Arab States, Cairo; Council for Scientific and Industrial Research (CSIR), Ghana; Iranian National Documentation Centre, Tehran; etc.

Author and co-author of six books; research papers and technical reports on documentation and information (about 100), on history and bibliography of science (25), and on other subjects (10).

S. L. PIOTROWSKI. Professor at the University of Warsaw; Chair of Astrophysics; Director, Astronomical Observatory, Warsaw University; Chairman, Committee on Astronomy, Polish Academy of Sciences; Member of the Polish Academy of Sciences, the Royal Astronomical Society, the American Astronomical Society.

Articles include: 'An Analytical Method for the Determination of the Intermediary Orbit of an Eclipsing Variable', *Astrophysical Journal*, No. 108, and Harvard Reprints, No. 312; 'The Planetary Problem of Radiative Transfer', *Acta Astronomica*, No. 4, 1947; 'Collision of Asteroids', *Astronomical Journal*, No. 57, 1952; 'Asymptotic Case of the Diffusion of Light through an Optically Thick Scattering Layer', *Acta Astronomica*, No. 6, 1956; 'Variations of Orbital Elements in Binary Systems with Mass Transfer', *Acta Astronomica*, No. 14, 1964; 'The Dimensions of Gaseous Rings in Close

Binary Systems', *Astrophysics and Space Science*, No. 8, 1970; 'The Dimensions of Circumstellar Discs in Binary Systems', *Acta Astronomica*, No. 25, 1975.

FREDERIC VESTER. Born 1925, Saarbrücken. Studied chemistry at the universities of Mainz and Paris. *Licence ès Sciences*, Sorbonne University, Paris, 1949; Doctoral thesis at the University of Hamburg. Research assistant, Institute for Experimental Cancer Research at Heidelberg, 1953–55; Post-doctoral Fellow, Medical School, Yale University, 1955–57; several research periods at the Nuclear Research Laboratories in Oakridge and Brookhaven, United States, and Cambridge, United Kingdom. Research assistant and lecturer in Biochemistry and Radio Chemistry, University of the Saarland, Saarbrücken, 1958–66; work with his own research group at the Max-Planck-Institut für Eiweiss- und Lederforschung, Munich, 1966–70. During this time, studies for a *Habilitation* degree at the University of Konstanz on cancerostatic plant proteins. Guest lecturer, Nuclear Research Centre, Karlsruhe School of Nuclear Technique, 1961–71. Since 1970, Privat-Dozent at the University of Konstanz; guest lecturer at several universities; Professor, Chair of Interdisciplinary Biology, University of Essen, 1975–76; President, Bavarian Adult Education Association, 1974–78; member of the scientific board of several governmental and private institutions; member of numerous scientific associations and publishing houses interested in new systems of education and communication. Until 1970, main fields: cancer research and biocybernetics. Since then, founder and head of the privately founded Study Group for Biology and Environment, a new type of institute for multi-disciplinary research, consulting and information; main fields: biocybernetics with application to the biology of learning and media, systems analysis and planning, environmental research, education and public health policy. Owner of several patents on cancerostatic proteins. Awards include: Adolf-Grimme Prize, 1974, for television film *Denken, Lernen, Vergessen* (Thinking, Learning, Forgetting); German Environmental Protection Medal, 1975; Josef-Hirt Prize, 1978, for outstanding contributions in the human sciences.

Publications on cancer research and biocybernetics as well as molecular biology, biophysics and research policy. Numerous scientific television series and radio broadcasts; production of educational films for schools. Most recent books include: *Das Überlebensprogramm*, Frankfurt, S. Fischer Verlag, 1975, with a Dutch translation; co-author with G. Henschel, *Krebs—Fehlgesteurtes Leben*, Munich, Deutscher Taschenbuch Verlag, 1977; *Denken, Lernen, Vergessen*, Stuttgart, Deutsche Verlagsanstalt, 1975, pocket edition, 1978, and translated into Dutch, Italian, Swedish, Spanish, Japanese (French in preparation); *Phänomen Stress*, Stuttgart, Deutsche Verlagsanstalt, 1976, pocket-book edition, 1978, translated into Dutch, Spanish, French, Norwegian (Swedish and Italian in preparation); *Ballungsgebiete in der Krise/Urban Systems in Crisis* (bilingual), Stuttgart, Deutsche Verlagsanstalt, 1976; *Unsere Welt—Ein Vernetztes System*, Stuttgart, Klett Cotta Verlag, 1978; co-author with H. Stern, G. Thielcke, R. Schreiber, *Rettet die Voegel—Wir Brauchen Sie*, Munich, Herbig Verlag, 1978; *Das Faule Ei des Kolumbus—Ein Energie-Bilderbuch*, Munich, Studiengruppe für Biologie und Umwelt, 1978.

GERALD JAMES WHITROW. Born 1912, in Kimmeridge, Dorset, England.

Educated at Christ's Hospital and Christ Church College, University of Oxford. Honours degree (first class) in mathematics, 1933; Harmsworth Senior Scholar, Merton College, Oxford, 1935–36. Research Lecturer, Christ Church College, Oxford, 1936–40; Scientific Officer, Ministry of Supply, 1940–45; on the staff of the Department of Mathematics, Imperial College, University of London, since 1945. At present, Professor of the History and Applications of Mathematics. Past President, British Society for the Philosophy of Science, British Society for the History of Science, British Society for the History of Mathematics, and First President of the International Society for the Study of Time; Vice-President, Royal Astronomical Society, 1966–67.

Author of numerous papers on relativity, cosmology, history and philosophy of science and mathematics. Publications include: *The Structure of the Universe*, 1949; *The Structure and Evolution of the Universe*, 1959; *The Natural Philosophy of Time*, 1961 (a new, completely revised and up-dated edition is now in course of publication by Oxford University Press); *What is Time?*, 1972; part author of *Atoms and the Universe*, 1956, 1962 and 1973; and *Rival Theories of Cosmology*, 1960. Editor of *Einstein, the Man and his Achievement*, 1967, and *Kant's Cosmogony*, 1970.

APPENDIX

Meeting on 'Time and the Sciences',
Royal Institution of Great Britain,
London, 4–6 February 1976

LIST OF PARTICIPANTS AND/OR AUTHORS OF ESSAYS[1]

Dr Ahmed M. Abou-Zeid, Professor of Sociology and Anthropology, Head of the Anthropology Department, Faculty of Arts, University of Alexandria (Egypt).

Ms Josefina Mena Abraham, Architect, Lisbon (Portugal).

Dr F. N'Sougan Agblemagnon, Laboratoire Africain de Coordination, de Recherche et d'Études Interdisciplinaires, 92360 Meudon-la-Forêt (France).

Professor Paul Fraisse, Laboratoire de Psychologie Expérimentale, Université René Descartes (Paris V), associated with the Centre National de la Recherche Scientifique, 28 rue Serpente, 75006 Paris (France).

Dr Jorge E. Hardoy, formerly with the Instituto Torcuato di Tella, Centro de Estudios Urbanos y Regionales, Buenos Aires (Argentina); University of Sussex (United Kingdom).

Professor Dr Mübeccel Kiray, Department of Social Sciences, Bosphorus University, Bebek P. K. 2, Istanbul (Turkey).

Dr John McHale, Director for Integrative Studies, School of Advanced Technology, State University of New York, Binghamton, New York, N.Y. 13901 (United States of America).

Dr A. Neelameghan, Head, Documentation, Research and Training Centre, Indian Statistical Institute, 112 Cross Road 11, Mallaswaren, Bangalore, 56003 (India).

---

1. Participants were associated with the institutions indicated at the time of the meeting (February 1976).

Professor S. Piotrowski, Astronomical Observatory, Warsaw University, Warsaw (Poland).

Privat-Dozent Dr Frederic Vester, Director, Studiengruppe für Biologie und Umwelt, Nussbaumstrasse 14, D 8000–München 2 (Federal Republic of Germany).

Mr G. J. Whitrow, Professor of History of Mathematics, University of London, Imperial College of Science and Technology, Exhibition Road, London SW7 (United Kingdom).

REPRESENTATIVES AND OBSERVERS

Non-governmental organizations

International Council of Scientific Unions (ICSU): Mr F. W. G. Baker, Secretary-General.

International Council for Philosophy and Humanistic Studies (ICPHS): Professor André Mercier, Secretary-General.

International Union of the History and Philosophy of Science (IUHPS): Dr Frank Greenaway, Secretary-General.

Other organizations

British Academy: Dr Stephen Körner, Department of Philosophy, University of Bristol (United Kingdom).

British Broadcasting Corporation: Mr Karl Sabbagh, Science and Features Television, BBC/TV, Kensington House, London (United Kingdom).

Royal Society: Professor J. M. Ziman, F.R.S., Department of Physics, University of Bristol (United Kingdom).

United Kingdom National Commission for Unesco.

MEMBERS OF UNESCO SECRETARIAT

Professor R. Habachi, Director, Philosophy Division.
Ms M. H. Vulcanesco, Philosophy Division.

[A.38] SS.78/XXII-3.A